Smart Innovation, Systems and Technologies

Volume 29

Series editors

Robert J. Howlett, KES International, Shoreham-by-Sea, UK
e-mail: rjhowlett@kesinternational.org

Lakhmi C. Jain, University of Canberra, Canberra, Australia
e-mail: Lakhmi.jain@unisa.edu.au

The Smart Innovation, Systems and Technologies book series encompasses the topics of knowledge, intelligence, innovation and sustainability. The aim of the series is to make available a platform for the publication of books on all aspects of single and multi-disciplinary research on these themes in order to make the latest results available in a readily-accessible form. Volumes on interdisciplinary research combining two or more of these areas is particularly sought.

The series covers systems and paradigms that employ knowledge and intelligence in a broad sense. Its scope is systems having embedded knowledge and intelligence, which may be applied to the solution of world problems in industry, the environment and the community. It also focusses on the knowledge-transfer methodologies and innovation strategies employed to make this happen effectively. The combination of intelligent systems tools and a broad range of applications introduces a need for a synergy of disciplines from science, technology, business and the humanities. The series will include conference proceedings, edited collections, monographs, handbooks, reference books, and other relevant types of book in areas of science and technology where smart systems and technologies can offer innovative solutions.

High quality content is an essential feature for all book proposals accepted for the series. It is expected that editors of all accepted volumes will ensure that contributions are subjected to an appropriate level of reviewing process and adhere to KES quality principles.

More information about this series at http://www.springer.com/series/8767

Jun Feng · Toyohide Watanabe

Index and Query Methods in Road Networks

 Springer

Jun Feng
Hohai University
Nanjing
China

Toyohide Watanabe
Nagoya Industrial Science
    Research Institute
Nagoya
Japan

ISSN 2190-3018
ISBN 978-3-319-34704-2
DOI 10.1007/978-3-319-10789-9

ISSN 2190-3026   (electronic)
ISBN 978-3-319-10789-9   (eBook)

Springer Cham Heidelberg New York Dordrecht London

Printed on acid-free paper

Springer is part of Springer Science+Business Media (www.springer.com)

Righteousness and affection

Preface

There has been an explosive growth of wireless communications technology, global positioning technology, and computer technology during the last decade. It is possible to use the spatial information to provide users with more services beyond now.

ITS uses advanced processing technology of spatial information, computer technology, control technology, electronic sensor technology, communications technology, and other means of transmission technologies to improve traditional traffic management system. It unifies people, vehicles, and roads, which can be real-time, accurate, and efficient traffic management and greatly decrease the traffic pressure. Currently, the actual investment using the ITS traffic monitoring system on the urban road network has the following steps:

1. traffic detectors are installed in each intersection to collect traffic flow information in real time;
2. communication equipment sends traffic flow information to the traffic control system in real time;
3. control system uses advanced mathematical model to optimize the signal control mode in each intersection.

Meanwhile, ITS can also use real-time vehicle information collected to monitor specific vehicle and support intelligent transportation services, such as:

1. analysis of a particular road traffic congestion in a particular time. For example, traffic monitoring system concerns about how many cars would pass Beijing Road between 7:00 and 8:00 during rush hour;
2. forecast of traffic flow to regulate traffic lights, then further control traffic flow and relieve traffic pressure based on the current traffic conditions. For example, prediction about how many vehicles would pass Beijing Road in the next 10 min.

Such services are based on the spatial-temporal query for a number of transportation vehicles which are moving objects. This book concerns the index and query techniques on road network and moving objects, which are limited to road network. Here, the road network of non-Euclidean space has its unique characteristics such that two moving objects may be very close in a straight line distance, but very far in road network; or two moving objects travel in different directions with small speed angle are close now, but they would be very far in a short time. So if you use index in two-dimensional Euclidean space to query moving objects on road network, the query will no longer have the superiority in efficiency and may even lead to incorrect query results. Therefore, we need to improve the index structure in order to obtain a suitable indexing method, explore the shortest path, and acquire nearest neighbor query and aggregation query methods under the new index structure.

Chapter 1 of this book introduces the present situation of intelligent traffic and index in road network, Chap. 2 introduces the relevant existing spatial indexing methods. Chapters 3–5 focus on several issues of road network and query, they involve: traffic road network models (see Chap. 3), index structures (see Chap. 4) and aggregate query methods (see Chap. 5). Finally, in Chap. 6, the book briefly describes the applications and the development of intelligent transportation in the future.

We started our research on spatio-temporal data management 15 years ago by chance when Jun Feng became a doctoral student of Prof. Toyohide Watanabe, who was supported by the Monbu-Kagaku-sho scholarship of the Ministry of Education, Science and Culture, Japan. And in the following years, we are constantly recruiting master and doctorial students in China and Japan to continue our research.

Many people have helped us in the preparation of this book. We would especially like to thank Zhonghua Zhu, Chunyan Lu, Jiamin Lu, Linyan Wu, Caihua Rui for their contributions to our research work. We would also like to thank Zhixian Tang, Zhenyu Sheng, Liming Xu, Yaqing Shi, Xiao Xu… for their careful and meticulous work during the writing and composing process.

Acknowledgment is also due to the National Science Foundation of China (No. 60673141 and No. 61370091) for partially supporting Jun's research reported here.

Last but not least, we would like to thank our families for their love, support, and patience.

Nanjing, China, April 2014   Jun Feng
                                                                        Toyohide Watanabe

Contents

1	**Introduction**	1
	1.1 Overview	1
	1.2 Road Network Modeling	2
	1.2.1 Non-Euclidean Feature of Road Networks	4
	1.2.2 Multi-levels Road Network	5
	1.3 Index Techniques in Road Network	6
	1.4 Query Methods in Road Network	6
	1.4.1 Precise Query Methods in Road Network	7
	1.4.2 Aggregate Query Methods in Road Network	7
	1.5 Cloud for Intelligent Transportation	8
	1.6 Summary	9
2	**Index Techniques**	11
	2.1 Binary-Tree Based Index Techniques	11
	2.1.1 kd-Tree	12
	2.1.2 K-D-B-Tree	13
	2.1.3 BSP-Tree	14
	2.1.4 Matsuyama's kd-Tree	14
	2.1.5 4d-Tree	15
	2.1.6 Skd-Tree	16
	2.2 B-Tree Based Index Techniques	18
	2.2.1 R-Tree	20
	2.2.2 R*-Tree	22
	2.2.3 R$^+$-Tree	24
	2.2.4 Hilbert R-Tree	24
	2.2.5 P-Tree	26
	2.3 Quad-Tree Based Structures	26
	2.3.1 Point Quad-Tree	27
	2.3.2 MX Quad-Tree	27
	2.3.3 PR Quad-Tree	30
	2.3.4 MX-CIF Quad-Tree	30

2.4   Cell Methods Based on Dynamic Hashing   33
      2.4.1   Grid File   33
      2.4.2   R-File   35
2.5   Spatial Objects Ordering......................................   36
      2.5.1   Z-Order Curve   37
      2.5.2   Hilbert Curve   38
2.6   Summary   38

3 Road Network Model.....................................   41
3.1   Map Information Model   41
      3.1.1   L-Model and T-Model   41
      3.1.2 M^2 Map Information Model   46
3.2   Multi-levels Model for Transportation Network   59
      3.2.1   Representation of Transportation Information   59
      3.2.2   Modeling of Road Network and Traffic Information   61
      3.2.3   Representation of Multi-levels
              of Transportation Network   64
3.3   Summary   69

4 Index in Road Network   71
4.1   R-TPR$^{\pm}$ Tree   72
      4.1.1   Introduction   72
      4.1.2   Road Connection Algorithms   73
      4.1.3   Framework and Query Method   74
      4.1.4   Evaluation   77
4.2   MOR-Tree   77
      4.2.1   Introduction   77
      4.2.2   Index Structure   79
      4.2.3   Algorithms for Operations of MOR-Tree   80
      4.2.4   Indexing Process for Two-Level Road Networks.......   82
      4.2.5   Evaluation   85
4.3   Sketch RR-Tree....................................   88
      4.3.1   Sketch and Sketch Index   88
      4.3.2   RR-Tree for Road Networks....................   91
      4.3.3   Structure of Sketch RR-Tree....................   91
      4.3.4   Operations on Sketch RR-Tree   92
      4.3.5   Evaluation   93
4.4   DynSketch   94
      4.4.1   Introduction   94
      4.4.2   Histogram   95
      4.4.3   Fitting Sketch................................   96
      4.4.4   Framework..................................   97
      4.4.5   Update of Buckets and Road Segments   99
      4.4.6   Algorithm of Search Using DynSketch..............   99
      4.4.7   Evaluation   101

4.5 Modified Histogram................................... 102
    4.5.1 Introduction 102
    4.5.2 Motivation 103
    4.5.3 Framework................................... 104
    4.5.4 Evaluation 105
4.6 Summary 106

5 Query in Road Network................................. 107
5.1 Nearest Neighbor Search on Road Network................ 108
    5.1.1 Introduction 108
    5.1.2 Framework of Cyclic Optimal Multi-step Method 108
    5.1.3 Cyclic Optimal Multi-step Algorithm................ 111
    5.1.4 Algorithm for Theoretical Analysis 113
    5.1.5 Evaluation 115
5.2 Continuous Nearest Neighbor Search on Road Network 117
    5.2.1 Introduction 117
    5.2.2 Road Network, Route and Computation Point......... 117
    5.2.3 Path Search Regions 118
    5.2.4 CNN-Search Approach.......................... 120
    5.2.5 Algorithm for Large Hierarchical Road Network....... 123
    5.2.6 Evaluation 126
5.3 Reverse Search Method of CNN 130
    5.3.1 Introduction 130
    5.3.2 Temporal Continuous Nearest Neighbor Search........ 130
    5.3.3 Algorithm Description 132
    5.3.4 Evaluation 133
5.4 Forecasting Aggregate Query on Road Network 135
    5.4.1 Introduction 135
    5.4.2 Exponential Smoothing 136
    5.4.3 Self-Adaptive Exponential Smoothing 139
    5.4.4 Transition Exponential Smoothing 143
5.5 Summary 146

6 The Trend of Development.............................. 147
6.1 Intelligent Transportation Cloud....................... 148
6.2 The Storage Techniques for Transportation Big Data 150
6.3 Challenges to Transportation Big Data Processing 152
6.4 Knowledge Discovery from Transportation Big Data 153
6.5 Summary 154

References.. 155

Chapter 1
Introduction

1.1 Overview

In recent years, with the rapid economic development, the fact that the number of vehicles grows rapidly leads to the great demand for urban transportation management. Although many departments of urban transportation have strengthened the construction of road networks management and have improved the efficiency of transportation systems, the relationship between supply and demand for transportation has not been balanced and many necessary facilities are still in short of supply. Thus, this phenomena causes traffic congestion and makes people difficult to travel. Today, traffic congestion has become a serious problem faced by major cities of the world.

Traffic congestion which has many problems in different aspects is difficult to deal with and it should be solved in various approaches. Currently, there are many methods adopted to solve the traffic congestion such as:

1. road-widening, which gives the reasonable planning of road infrastructure. However, the slow pace of road-widening cannot catch up with the growth rate of vehicles;
2. request for congestion charge in the city center, which uses economic approaches to reduce the number of vehicles;
3. to use radio to provide real-time traffic information, which indicates the travel routes in advance, but the accuracy of these information is inadequate;
4. to use the strategy which chooses odd or even license plate number in turn to limit vehicles traveling;
5. to improve the rate of public transportation and create a fast and comfortable environment of public transportation.

However, these methods cannot solve the problems of traffic congestion fundamentally. So in the beginning of 1990s, the United States, Japan and Europe began to adopt information technology to solve this problem and they proposed the intelligent transportation systems (ITS) conception. ITS uses advanced information

© Springer International Publishing Switzerland 2015
J. Feng and T. Watanabe, *Index and Query Methods in Road Networks*,
Smart Innovation, Systems and Technologies 29,
DOI 10.1007/978-3-319-10789-9_1

technology, computer technology, control technology, electronic sensor technology and communication transmission technology to transform the traditional traffic management systems, which unifies people, vehicles and roads. Transportation can be managed accurately and efficiently, which greatly reduces the traffic pressure.

As the development of wireless communication technology, global positioning technology and computer technology, it is possible to use the spatial information to provide users with new services (called Location-Based Service, LBS) such as the vehicle monitoring, dynamic route search, mobile e-commerce, which greatly promote the development and applications of intelligent transportation. As an important location-related application, MOD (Moving Objects Databases) technology has become a research hotspot, and it is the database which represents and manages the position of moving objects and related information [1]. In the real world, taking into account the different mobile objects and their applications, movement of the object can be divided into non-limited movement (such as the movement of the submarine in the ocean), restricted movement (such as moving pedestrian) and the movement based on spatial networks (such as the car or train moving in traffic network) [2]. Among them, the movement based on spatial networks is the most general. Especially, with the continuous development of urban transportation systems, it has become a serious problem to achieve the real-time and efficient management of the urban traffic network.

Mobile services for urban traffic moving objects are mostly based on current and predicted location information. In the large-scale transportation network or a large number of moving objects (such as urban transportation network and vehicles on it), spatial-temporal information retrieval efficiency is the key to meet real-time requirements of the location-based services. To solve the efficiency problem of information access in road network, the most direct way is to study and propose an efficient spatial index structure based on spatial network to organize the physical storage of information. However, the problem is not isolated; the research on aggregation index methods over data streams in road network involves the following questions: traffic road network modeling (see Chap. 3), index structure (see Chap. 4), query methods of moving objects (see Chap. 5) and some applications and development trend (see Chap. 6).

1.2 Road Network Modeling

As road network is formed under natural conditions, social conditions and local construction conditions in order to meet the various requirements of transportation, it has no uniform format of representation. Figure 1.1a shows us a real road network and Fig. 1.1b is a model of this road network. We use $Road_i$ to represent road segment and V_i to represent intersection. At the intersection V_i, vehicles can either move along the original path of the original direction, or change the direction and travel on other roads. In $Road_i$, there are several inflection points n_i. At the inflection point n_i, vehicles can only move along the original path, but they can change the directions.

(a)

(b)

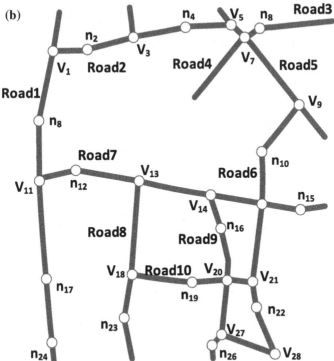

Fig. 1.1 Road network modeling. **a** Real road network. **b** Modeling of road network

For example, $Road_7$ has two intersections (V_{11} and V_{13}) and one inflection point (n_{12}). If a vehicle is on $Road_7$ and moves to intersection 13 (V_{13}), it can either move along the original path of the original direction, or change the direction and travel on $Road_8$. While a vehicle can only move along the original path, but it can change the direction at n_{12}. We can see that intersections and inflection points are different and moving objects are limited by road network. A typical problem is how to deal with non-Euclidean feature of road networks.

1.2.1 Non-Euclidean Feature of Road Networks

In road networks, the movement of moving objects is limited by the structure of road networks. So the model of road networks is a typical non-Euclidean space model. As shown in Fig. 1.2, A_1 and A_2 represent the gas station respectively and a car is moving on the roadway. Consider this situation: this car wants to find the nearest gas station. In Euclidean space, d_0 is the distance from the car to A_1 and d_0' is the distance from the car to A_2. As $d_0 < d_0'$, A_1 is the nearest target. While in non-Euclidean space (road network), $d_1+d_2+d_3$ is the distance from the car to A_1 and d_1+d_2 is the distance from the car to A_2. As $d_1+d_2 < d_1+d_2+d_3$, A_2 is the nearest target. The distance from the car to the gas station is not computed with the coordinates of these two locations (represented by black dotted line), but is based on the path length (solid line). We can see that the situation in non-Euclidean space is obviously different from that in Euclidean space. When we search for the targets in road network, non-Euclidean space is important in consideration.

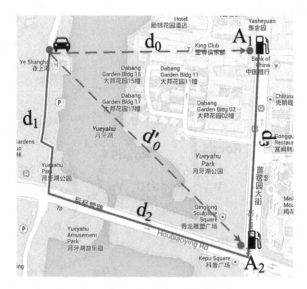

Fig. 1.2 Example of non-Euclidean space in road network

1.2.2 Multi-levels Road Network

It is known to us that maps are usually divided into different parts according to administrative areas. As shown in Fig. 1.3, map has many levels such as country level, prefecture level, city level and so on, which forms a tree structure. It is the same as road network which is also divided into different sub-networks according to countries, prefectures, cities and so on. We call this multi-levels transportation network. Our queries may be in different levels of road network. When we want to search for a specific location like a gas station, we prefer to execute the query in a small region like a street. While, if we want to gather summarized information, we would rather execute the query in a large region like a prefecture.

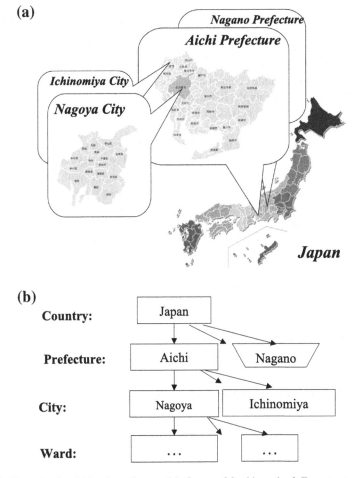

Fig. 1.3 Example of multi-levels road network in Japan. **a** Map hierarchy. **b** Tree structure for map hierarchy

It is noticed that road networks on different scales are independent and they are created and maintained respectively on different levels. We still need to keep information consistent for multi-levels road network and build relationships between road networks on different scales. Modeling methods are used to represent road network and can also process the problems in multi-levels transportation network. Such a M^2 map information model (to be mentioned in Sect. 3.2) can ensure that maps are created and maintained respectively on different scales and that information consistency can be remained. It also builds relationship between maps on different scales.

1.3 Index Techniques in Road Network

Index techniques are usually used to improve the efficiency of query. However, distance between source and target in road network is not computed with the coordinates (spatial data) of these two locations. It is computed based on the path length (geographical relation) between them. Since road network belongs to a non-Euclidean space, spatial index cannot be used directly, so we need other methods to index road network. For example, RR-tree makes full use of advantages of R-tree and it can index vehicles in road network efficiently. MOR-tree can index road network on different scales.

To index road networks, there is another important problem we cannot ignore. We should consider the big difference between urban and rural economy which makes the density of vehicles vary widely in the urban and the rural. With the development of city scale, even in the same city at the same time there is a big difference in the distribution of moving objects. Non-uniform distribution of moving objects would cause many problems. For example, query response time difference among different areas would lead to difficulties in decision-making. In addition, the same query methods almost have the same relative errors. While, more objects would lead to more absolute errors. Then, the quality of query cannot be ensured, which would impede the improvement of traffic situation.

To solve non-uniform distribution problems, we need an intelligent region-dividing method to ensure the efficiency of query in different areas and to improve the quality of query (referred to Sect. 4.4).

1.4 Query Methods in Road Network

There are many daily applications in road network. They are described as follows:

- Road-widening, which gives the reasonable planning of road infrastructure. However, the slow pace of road-widening cannot catch up with the growth rate of vehicles.

- Request for congestion charge in the city center, which uses economic approaches to reduce the number of vehicles.
- To use radio to provide real-time traffic information, which indicates the travel routes in advance, but the accuracy of these information is inadequate.
- To use the strategy which chooses odd or even license plate number in turn to limit vehicles traveling.
- To improve the rate of public transportation and create a fast and comfortable environment of public transportation.

All above applications require query or search requests, but these query requests are not the same. In the first three applications such as to find a hotel, to look for a gas station or to search some people, we have to know the exact location of each target; Otherwise we cannot arrive to destinations. In the last two applications, we only need to know summarized information of each road segment rather than any specifics. So we can divide these applications into two categories: precise query and aggregate query.

1.4.1 Precise Query Methods in Road Network

There are three types of precise queries discussed in this book: nearest neighbor query(NN), continuous nearest neighbor query(CNN) and continuous k nearest neighbor query(CKNN) (to be mentioned in Sects. 5.1–5.3).

- NN: find the nearest objects for a static query object. The number of results can be one or more.
- CNN: find the nearest objects for a moving query object continuously.
- CKNN: find k nearest objects for a moving query object continuously.

Each type of queries corresponds to some applications. These queries belong to precise queries which would get exact location in road network and they are used widely in ITS. Non-Euclidean space (to be mentioned in Sect. 1.2) is the most serious problem in these queries and we can use COMS method to solve this problem.

1.4.2 Aggregate Query Methods in Road Network

Aggregate query aims at obtaining summarized information such as vehicles counts. In this situation, moving objects' snapshots are gathered continuously. Distinct counting problem and non-uniform distribution problem are prominent for aggregate query. For example, when we execute aggregate query for specific road segments during a period of time, some vehicles may be computed multiple times during the query period of time. On the other hand, when vehicles density of some road segments is larger than that of other road segments, it is difficult to get query results by using the same aggregate method.

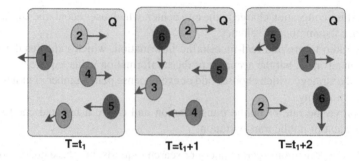

Fig. 1.4 Example of distinct counting problem

As previously mentioned, in many applications of aggregate query, we usually need statistic information of road network such as vehicles counting (e.g., how many vehicles have passed through Tiananmen Square from 8:00 to 9:00 this morning?). As shown in Fig. 1.4, Q is the query area and there are some moving objects in it. At time t_1, there are 5 objects in area Q. At time t_1+1, there are also 5 objects in area Q, while some of these objects are the same as those at time t_1. At time t_1+2, there are 4 objects and some of these objects are still the same as those at times t_1 and t_1+1. If we want to know how many objects emerged from t_1 to t_1+2 inside area Q, some objects would be computed multiple times such as object1, object2, object3. This is called distinct counting problem.

Aggregate query would gather massive information and process a large number of moving objects. So it usually takes a lot of time. If we want to speed up query process, it is better to reduce the number of counts. We require techniques which can solve the distinct counting problem to improve the efficiency of aggregate query. In the following chapters, we can use Sketch-based methods such as Sketch-RR tree (to be mentioned in Sect. 4.3), DynSketch (to be mentioned in Sect. 4.4), MH (to be mentioned in Sect. 4.5) methods to solve above problems in aggregate query.

1.5 Cloud for Intelligent Transportation

Cloud computing technology which has been developed in recent years is a new type of computing patterns. Cloud computing embodies a new concept of information services. Cloud computing is the key technique of solving the problem of massive data with its automated computer resource scheduling, deployment of high-speed information and excellent scalability. As an emerging computing and business model, cloud computing is accelerating the processes of transportation information service and information industry. Rapid development of cloud computing in the field of intelligent transportation applications has positive significance to improve the integrated information processing capacity of the cities and promote the upgrading of the industrial optimization and the structures. At the same time, cloud computing is promoting the transformation of the mode economic development, which has a broad market prospect.

Intelligent transportation cloud is based on the data streams of the road networks. Intelligent transportation cloud uses the excellent data processing capabilities of cloud computing to improve the performance of the intelligent transportation systems (ITS) as well as its scalability, reliability and cost benefits, and it provides strong support for intelligent transportation system.

1.6 Summary

Intelligent Transportation System (ITS) is based on the increasing demands of the transportation development. It integrates information, communications, computers and other technologies, and applies them in the field of transportation to build an integrated system of people, roads and vehicles by utilizing advanced data communication technologies. Roadways play the role as a carrier which is used to limit the activities of people and vehicles. Technologies in road network contribute to establish a large, full-functioning, realtime, accurate and efficient transportation management system. With the development of ITS, researches on the road network will get a wide range of industry and academic attention. This chapter briefly introduced the road networks modeling, index and query for moving objects and some typical problems in the applications of road networks.

Chapter 2
Index Techniques

The efficiency of data access and storage is a key factor that affects the quality of data service, and it can be significantly improved by effective index mechanism. Data index is a structure used to organize data records and describe the location information of data in physical storage medium. Index techniques can help us access record set through multiple ways and effectively support many kinds of queries. There are two kinds of index method in traditional database system: the first one is tree (e.g., B^+-tree or B-tree) based index, and the second is hash based index. Search engine often uses inverted file as its index method. Spatial and temporal data indexes (e.g. R-tree and its variants, K-D-tree and its variants, and space filling curves) are mainly extended from traditional database index. This book centers on index and query techniques in road network. As a complicated data structure, road network not only contains static spatial data, such as roads, lakes, and buildings, but also includes dynamical spatio-temporal data, e.g., the location information of mobile objects. So, to index road network we must use various types of current index techniques holistically. In order to analyze the index methods in road network well, in this chapter, we briefly examine the typical indexes proposed in the literature and present a basic description on them.

2.1 Binary-Tree Based Index Techniques

The binary search tree is a basic data structure for representing data items whose index values are arranged in some linear order. The idea of repetitively partitioning a data space has been adopted and generalized in many sophisticated indexes. In this section, we will examine indexes originated from the basic structure and concept of binary search trees.

Finally, we would like to further emphasize that solutions to all above mentioned issues require close and efficient collaborations between the computer scientists and the application developers. High performance index techniques can only be developed with a through understanding of the usage of spatial data, including the

© Springer International Publishing Switzerland 2015
J. Feng and T. Watanabe, *Index and Query Methods in Road Networks*,
Smart Innovation, Systems and Technologies 29,
DOI 10.1007/978-3-319-10789-9_2

access patterns and the post processing after data brought into memory. At the same time, application developers may be able to provide certain services or tune their algorithms to avoid some of the limitations of underlying indexing mechanism.

2.1.1 kd-Tree

A kd-tree [3] (short for multi-dimensinal binary search tree) is a space-partitioning data structure for organizing points in a multi-dimensional space, which was introduced by Bentley in 1975. The kd-tree is a natural generalization of the well-known binary search tree to handle the case of a single record having multiple keys. Differed from the binary tree, a node in kd-tree (Fig. 2.1) is a k-dimensional point and serves two purposes: representing an actual data point and giving the direction of a search. In every level, there is a discriminator whose value is between 0 and $k-1$ inclusive, which indicates the key on which the branching depends. A node P has two children, a left son $LOSON(P)$ and a right son $HISON(P)$. If the discriminator value of node P is the jth key (attribute), the jth key of any node in the $HISON(P)$ is greater than or equal to that of node P. This feature enables the range along each dimension to be

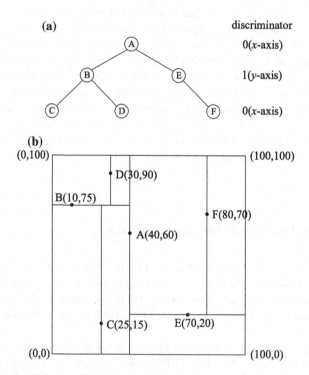

Fig. 2.1 Example of kd-tree. **a** Tree structure. **b** Planar structure

defined during a tree traversal such that the ranges are smaller in the lower levels of the tree. To keep this property, deletion will probably cause successive replacements. In order to reduce the cost of deletion, Bentley proposed a non-homogeneous kd-tree in 1979 [4]. Unlike a homogeneous index, a non-homogeneous index does not store data in the internal nodes and its internal nodes are used only as directories. The kd-tree has been the subject of intensive research over the past decades. Many variants have been proposed in the literature to improve the performance of the kd-tree with respect to issues such as clustering, searching, storage efficiency and balancing.

2.1.2 K-D-B-Tree

To improve the paging capability of the kd-tree, Robinson proposed the K-D-B tree [5] which combines the properties of kd-tree [3] and B-tree [6, 7].The K-D-B tree consists of two basic parts: region pages (internal node) and point pages (leaf node) (see Fig. 2.2). While point pages contain object identifiers, region pages store the descriptions of subspaces in which the data points are stored and the pointers to descendant pages. In K-D-B tree, these subspaces are explicitly stored in a region page. These subspaces such as $S11$, $S12$, $S13$, are pairwise disjoint and together they span the rectangular subspace of the current region page (e.g., $S1$), a subspace in the parent region page.

When inserting a new point into a full point page, a split will happen. The point page is split so that the two resultant point pages will contain almost the same number of data points. Note that the spit of a point page requires an extra entry of a new point page. This entry will be inserted into the parent region page. Therefore, the split of a point page may cause the parent region page to split as well, which may further ripple all the way to the root. Thus the tree is always perfectly height-balanced.

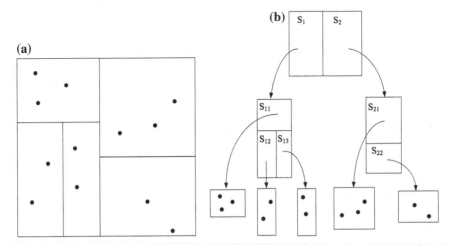

Fig. 2.2 Example of K-D-B tree. **a** Area devision. **b** Tree structure

When a region page is split, the entries are partitioned into two groups such that both have almost the same number of entries. A hyperplane is used to split the space of a region page into two subspaces and this hyperplane may cut across the subspaces of some entries. Consequently, the subspaces that intersect with the splitting hyperplane must also be split so that the new subspaces are totally contained in the resultant region pages. If the constraint of splitting a region page into two, containing the same number of entries is not enforced, then downward propagation of split may be avoided. The choice of the dimension for splitting and the splitting point would be chosen so that both resultant pages have almost the same number of entries and the number of splitting is minimized.

The upward propagation of a split would not cause the underflow of pages, but the downward propagation is detrimental to storage efficiency because a page may contain less than the usual threshold, typically half of the page capacity. To avoid unacceptabe low storage utilization, local reorganization can be performed. For example, two or more pages whose data space forms a rectangular space and they having the same parent can be merged followed by a re-split if the resultant page overflows.

2.1.3 BSP-Tree

A Binary Space Partitioning tree (or BSP-tree) [8, 9] is a data structure that is used to organize objects within a space. Like kd-trees, BSP-trees are binary trees that represent a recursive subdivision of the universe into subspaces by means of $(d-1)$-dimensional hyperplanes. Each subspace is subdivided independently according to its history and other subspaces. The choice of the partitioning hyperplanes depends on the distribution of the data objects in a given subspace. The decomposition usually continues until the number of objects in each subspace is below a given threshold. The resulting partition of the universe can be represented by a BSP-tree in which each hyperplane corresponds to an interior node of the tree and each subspace corresponds to a leaf. Each leaf stores references to those objects that are contained in the corresponding subspace.

Binary space partitioning was developed in the context of 3D computer graphics, where the structure of a BSP-tree allows spatial information about the objects in a scene that is useful in rendering, such as their ordering from front-to-back with respect to a viewer at a given location, to be accessed rapidly. Other applications include performing geometrical operations with shapes (constructive solid geometry) in CAD, collision detection in robotics and 3D video games, ray tracing and other computer applications that involve handling of complex spatial scenes.

2.1.4 Matsuyama's kd-Tree

While most kd-trees are proposed as point access methods, the kd-tree proposed by Matsuyama et al. [10] is designed for two-dimensional non-zero sized spatial objects by supporting duplications of objects. The directory is a kd-tree, and for each leaf node, a data page is associated. A data page contains the identifiers of objects which are partially or totally included in its data space. Objects that overlap multiple un-partitioned data space are duplicated in respective data pages.

Matsuyama's kd-tree is searched like a conventional kd-tree. However, to insert an object, the object identifier needs to be inserted into all the pages with subspaces that intersect with the data object. It is quite common that object identifiers may be duplicated in more than one page, particularly when the sizes of objects are large. Whenever a page overflows, the page is split with a partition being introduced along the longer side of the rectangle. The subspace is partitioned into two subspaces and the two new pages contain all objects that intersect with their subspace.

To delete an object, it is necessary to search all leaf nodes with subspaces that intersect with the data object and delete all identifiers referring to the data objects. If the deletion of an object causes a page to be empty, the corresponding leaf node should be marked NIL. To simplify the deletion algorithm, the underflowed data pages do not need to be merged.

Matsuyama's kd-tree is one of the earlier indexing structures adopting the object duplication approach. Such an index is not suitable for indexing large objects as the overhead of redundant storage can be very high.

2.1.5 4d-Tree

The kd-tree can be used to index two-dimensional rectangular objects by mapping the objects into points in a 4-dimensional space. Each two-dimensional rectangular described by (x_1, y_1) and (x_2, y_2), is treated as a four attribute tuple (x_1, x_2, y_1, y_2). The discriminator is used cyclically and the nodes at the same level use the same discriminator. In [11], the issues involved in mapping the data structure onto pages in secondary memory were not addressed. The same approach for the K-D-B tree [5] was suggested by Banerjee and Kim [12]. The structure is known as the 4d-tree.

At each node of the 4d-tree, a discriminator (x_1, x_2, y_1, y_2), discriminator value and pointers to two child nodes are stored. A two-dimensional subspace is associated with each node and as the tree is traversed during query, starting from the root, these subspaces are successively pruned. Let the query region be (qx_1, qx_2, qy_1, qy_2). Then, at each internal node, one of the conditions, $x_1 \leq qx_2, x_2 \geq qx_1, y_1 \leq qy_2$ or $y_2 \geq qy_1$, has to be used depending on the discriminator stored in that node in order to determine whether both subtrees or only one of the subtrees need to be searched.

The important part in the search algorithm is the determination of the subspaces that bound the objects in the LO (*left*) and HI (*right*) subtrees. Traversal starts at the root with the map as the associated space. Assume that the left discriminator is X_1, the LO subtree contains objects whose X_1 coordinate is less than the discriminator value, and the HI subtree contains objects whose X_1 coordinate is greater than the discriminator value. The X_1 values of the HI subspace are bounded below by the discriminator value and this fact can be used to reduce the subspace associated with

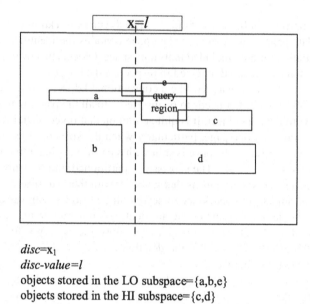

$disc$=x$_1$
$disc$-$value$=l
objects stored in the LO subspace={a,b,e}
objects stored in the HI subspace={c,d}

Fig. 2.3 A 4d-tree objects distribution

the HI subspace. For example, to search for objects that overlap a given object with X_2 less than l (discriminator value) in Fig. 2.3, we can conclude that the right subtree does not contain any objects that will intersect with the given object. However, it is not possible to reduce the size of the LO subspace. Suppose the original map space is (x_1, x_2, y_1, y_2). Then the LO subspace is the same as that of the root node while the HI subspace is $(disc\_value, x_2, y_1, y_2)$. The problem is that the X_2 values of rectangles in the left subspace may fall on the right subspace, and there is no information about extent to which they overlap. At the next level, the HI subspace remains unchanged, but for the LO subspace X_2 is bounded by the current discriminator value. Hence, it is common that both subtrees of a node need to be searched. The major problem associated with the 4d-tree is its intersection search, which can cost a lot due to the need for traversal of both subtrees when a query region lies in a subspace that cannot be bounded tightly using the discriminator values.

2.1.6 Skd-Tree

Ooi et al. [13, 14] developed an indexing structure called the spatial kd-tree (the Skd-tree) in an attempt to avoid object duplication and object mapping. At each node of a kd-tree, a value (the discriminator value) is chosen in one of the dimensions to partition a k-dimensional space into two subspaces. The two resultant subspaces, *HISON* and *LOSON*, normally have almost the same number of data objects. Point

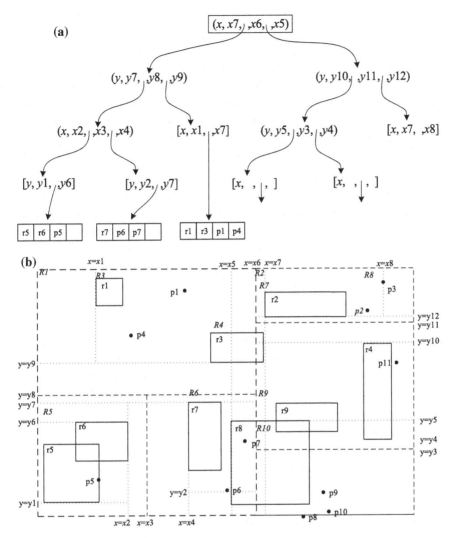

Fig. 2.4 Example of a spatial kd-tree. **a** 2-D directory of the skd tree. **b** 2-D space coordinate representation

objects are totally included in one of the two resultant subspaces, but non-zero sized objects may extend over to the other subspace. To avoid the division of objects and the duplication of identifiers in several subspaces, and yet to be able to retrieve all the wanted objects, literature [14] introduced a virtual subspace for each original subspace such that all objects are totally included in one of the two virtual subspaces. With this method, the placement of an object in a subspace is based solely upon the value of its centroid.

One additional value for each subspace is stored: the maximum (MAX_{LOSON}) of the objects in the $LOSON$ subspace, and the minimum (MIN_{HISON}) of the objects in the HISON subspace, along the dimension defined by the discriminator. The structure of an internal node of the Skd-tree consists of two child pointers, a discriminator (0 to $k - 1$ for a k-dimensional space), a discriminator-value, (MAX_{LOSON}) and (MIN_{HISON}) along the dimension specified by discriminator. The maximum range value of $LOSON(MAX_{LOSON})$ is the nearest virtual line that bounds the data objects whose centroids are in the $LOSON$ subspace, and the minimum range value of $HISON(MIN_{HISON})$ is the nearest virtual line that bounds the data objects whose centroids are in the $HISON$ subspace.

Leaf nodes contain min-range and max-range (in place of MAX_{LOSON} and MIN_{HISON} of an internal node respectively), describing the minimum and maximum values of objects in the data page along the dimension specified by bound, and a pointer to the secondary page which contains the object bounding rectangles and identifiers. The minimum and maximum values could be kept for k-dimensions. However, for storage efficiency, the range along one dimension that results in the smallest bounding rectangle is chosen. Figure 2.4a, b show the structure of a two-dimensioned Skd-tree and illustrate the virtual boundary (dotted line), MAX_{LOSON} or MIN_{HISON} of each resultant subspace.

An implicit rectangular space is associated with each node and it is materialized during traversal. This rectangle is tested against the query region, and the subtree is examined if they intersect. Since the virtual boundary may sometimes bound the objects tighter than the partitioning line, the intersection search takes advantage of the existing virtual boundary to prune the search space efficiently. To further exploit the virtual boundaries, containment search which retrieves all spatial objects contained in a given query rectangle was proposed. During tree traversal, the algorithm always selects the boundaries that yield smaller search space. The direct support of containment search is useful to operators like within and contain. The search rapidly eliminates all objects that are not totally contained in the query region.

2.2 B-Tree Based Index Techniques

In computer science, a B-tree [6, 7] is a tree data structure that keeps data sorted and allows searches, sequential access, insertions, and deletions in logarithmic time. The B-tree is a generalization of a binary search tree in which a node can have more than two children. Unlike self-balancing binary search trees, the B-tree is optimized for systems that read and write large blocks of data. It is commonly used in databases and file systems [15].

In B-trees, internal (non-leaf) nodes can have a variable number of child nodes within some predefined range. When data are inserted or removed from a node, its number of child nodes changes. In order to maintain the predefined range, internal nodes may be merged or split. Because a range of child nodes is permitted, B-trees do not need re-balancing as frequently as other self-balancing search trees, but may

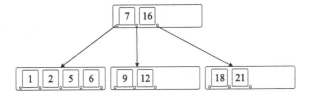

Fig. 2.5 A B-tree of order 2 or order 5

waste some space, since nodes are not entirely full. The lower and upper bounds on the number of child nodes are typically fixed for a particular implementation. For example, in a 2–3 B-tree (often simply referred to as a 2–3 tree), each internal node may have only two or three child nodes.

Each internal node in a B-tree will contain a number of keys. In general, each node in a B-tree whose order is d contains at most $2d$ keys and $2d + 1$ pointers, as shown in Fig. 2.5. Actually, the number of keys may vary from node to node, but each must have at least d keys and $d + 1$ pointers. As a result, each node is at least 1/2 full. The keys act as separation values which divide its subtrees. For example, if an internal node has three child nodes (or subtrees) then it must have 2 keys: $a1$ and $a2$. All keys in the leftmost subtree will be smaller than $a1$, all keys in the middle subtree will be between $a1$ and $a2$, and all keys in the rightmost subtree will be greater than $a2$.

Usually, the number of keys is chosen to vary between d and $2d$. In practice, the keys take up the most space in a node. If an internal node has $2d$ keys, adding a key to that node can be accomplished by splitting $2d$ key nodes into d key nodes and adding the key to the parent node. Each split node has the required minimum number of keys. Similarly, if an internal node and its neighbor each have d keys, then a key may be deleted from the internal node by combining with its neighbor. Deleting the key would make the internal node have $d - 1$ keys; and merging the neighbor would add d keys and one more key brought down from the neighbor parent.

The number of branches (or child nodes) from a node will be one more than the number of keys stored in the node. In a 2–3 B-tree, the internal nodes will store either one key (with two child nodes) or two keys (with three child nodes). A B-tree is sometimes described with the parameters from $(d + 1)$ to $(2d + 1)$ or simply with the highest branching order $(2d + 1)$.

A B-tree is kept balanced by requiring that all leaf nodes are at the same depth. This depth will increase slowly as elements are added to the tree, but an increase in the overall depth is infrequent.

B-trees have substantial advantages over alternative implementations when accessing the data of a node greatly exceeds the time spent processing these data, because the cost of accessing the node may be amortized over multiple operations within the node. This usually occurs when the node data are in secondary storage such as disk drives. By maximizing the number of child nodes within each internal node, the height of the tree decreases and the number of expensive node accesses is reduced. In addition, re-balancing of the tree occurs less often. The maximum number of child nodes depends on the information which must be stored for each child node and the

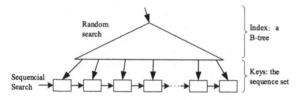

Fig. 2.6 A B^+-tree with separate index and key parts

size of a full disk block or an analogous size in secondary storage. While 2–3 B-trees are easier to explain, practical B-trees using secondary storage need a large number of child nodes to improve performance.

The term B-tree may refer to a specific design or may refer to a general class of designs. In the narrow sense, a B-tree stores keys in its internal nodes but need not store those keys in the records at the leaves. The general class includes variants such as the B*-tree and B^+-tree.

Perhaps the most misused term in B-tree literature is B*-tree. In fact, Knuth defines a B*-tree [16] to be a B-tree in which each node is at least 2/3 full (instead of just 1/2 full). B*-tree insertion employs a local redistribution scheme to delay splitting until two sibling nodes are full. Then the two nodes are divided into three, each 2/3 full. This scheme guarantees that storage utilization is at least 66 %, while requiring only moderate adjustment of the maintenance algorithms. It should be pointed out that increasing storage utilization has the side effect of speeding up the search since the height of the resulting tree is smaller.

In a B^+-tree, all keys reside in the leaves. The upper levels, which are organized as a B-tree, consist only of an index, a road map to enable rapid location of the index and key parts. Figure 2.6 shows the logical separation of the index and key parts. Naturally, index nodes and leaf nodes may have different formats or even different sizes. In particular, leaf nodes are usually linked together left-to-right, as shown in Fig. 2.6. The linked list of leaves is referred to as the sequence set. Sequence set links allow easy sequential processing.

2.2.1 R-Tree

R-tree [17] is a multi-dimensional generalization of the B-tree, that preserves height-balance. Like the B-tree, node splitting and merging are required for inserting and deleting objects. The R-tree has received a great deal of attention due to its well defined structure and the fact that it is one of the earliest proposed tree structures for indexing non-zero sized spatial object. Many papers have used the R-tree as a model to measure the performance of their structures.

An entry in a leaf node consists of an object-identifier of the data object and a k-dimensional bounding rectangle which bounds its data objects. In a non-leaf node,

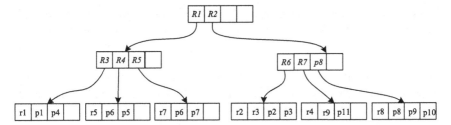

Fig. 2.7 Directory of a R-tree

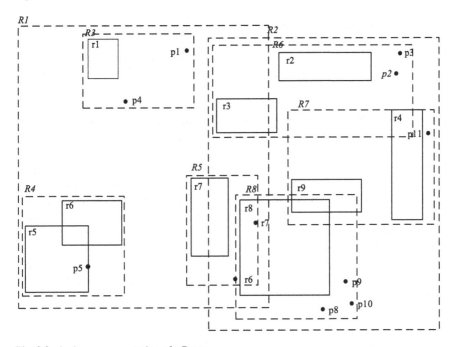

Fig. 2.8 A planar representation of a R-tree

an entry contains a child-pointer pointing to a lower level node in the R-tree and a bounding rectangle covering all the rectangles in the lower nodes in the subtree. Figures 2.7 and 2.8 illustrate the structure of an R-tree and its planar representation respectively.

In order to locate all objects which intersect a query rectangle, the search algorithm descends the tree from the root. The algorithm recursively traverses down the subtrees of bounding rectangles that intersect the query rectangle. When a leaf node is reached, bounding rectangles are tested against the query results and their objects are fetched for testing if they intersect the query rectangle.

To insert an object, the tree is traversed and all the rectangles in the current non-leaf node are examined. The constraint of least coverage is employed to insert an object: the rectangle that needs least enlargement to enclose the new object is selected and the one with the smallest area is chosen if more than one rectangle meets the first criterion. The nodes in the subtree indexed by the selected entry are examined recursively. Once a leaf node is obtained, a straightforward insertion is made if the leaf node is not full. However, the leaf node needs splitting if it overflows after the insertion is made. For each node that is traversed, the covering rectangle in the parent is readjusted to tightly bound the entries in the node. For a new split node, an entry with a covering rectangle that is large enough to cover all the entries in the new node is inserted in the parent node if there is room in the parent node. Otherwise, the parent node will be split and the process may propagate to the root.

To delete an object, the tree is traversed and each entry of a non-leaf node is checked to determine if the object overlaps its covering rectangle. For each entry, the entries in the child node are examined recursively. The deletion of an object may cause the leaf node to underflow. In this case, the node needs deleting and all the remaining entries of that node are reinserted from the root. Similar to the node splitting, the deletion of an entry may cause further deletion of nodes in the upper levels. Thus, entries belonging to a deleted ith level node must be reinserted into the nodes in the ith level of the tree. Deletion of an object may change the bounding rectangle of entries in the ancestor nodes. Hence readjustment of these entries is required.

In searching, the decision whether to visit a subtree depends on whether the covering rectangle overlaps the query region. It is quite common for several covering rectangles in an internal node which overlap the query rectangle, resulting in the traversal of several subtrees. Therefore, the minimization of overlaps of covering rectangles as well as the coverage of these rectangles is of primary importance in constructing the R-tree.

The heuristic optimization criterion used in the R-tree is the minimization of the area of internal nodes covering rectangles. In [17], splitting algorithms with exponential, quadratic and linear cost were discussed. Among them, the exponential algorithm can find the optimal solution, but the algorithm time complexity is highquadratic and linear algorithms time complexity is low and can get sub-optimal solution. The quadratic algorithm searches the pair of rectangles that is the worst combination to have in the same node, and puts them as initial objects into the two new groups. It then searches the entry which has the strongest preference for one of the groups (in terms of area increase) and assigns the object to this group until all objects are assigned (satisfying the minimum fill). The linear algorithm chooses the first two objects based on the separation between the objects in relation to the width of the entire group along the same dimension.

2.2.2 R*-Tree

Minimization of both coverage and overlaps is crucial to the performance of the R-tree. It is however impossible to minimize coverage and overlaps at the same time. A balancing criterion must be found so that the near optimal of both minimization can produce the best result. An additional optimization objective put forward in [18] is the margin of the covering rectangles. Squarish covering rectangles are preferred. Based on the fact that clustering rectangles with little variance of the lengths of the edges tend to reduce the area of the clusters covering rectangle, the criterion that ensures the quadratic covering rectangles is used in the insertion and splitting algorithms of the improved R-tree, called the R*-tree.

In the leaf nodes of the R*-tree, a new record is inserted into the page whose entry covering rectangle, if enlarged, has the least overlap with other covering rectangles. A tie is resolved by choosing the entry whose rectangle needs the least area enlargement. However, in the internal nodes, an entry whose covering rectangle needs the least area enlargement is chosen to include the new record, and a tie is resolved by choosing the entry with the smallest resultant area. The improvement is particularly significant when both the query rectangles and data rectangles are small, and when the data is non-uniformly distributed. In the R*-tree splitting algorithm, along each axis, the entries are sorted by the lower value, and also sorted by the upper value of the entry rectangles. For each sort, $M - 2m + 2$ [1] distributions of splits are considered, where in kth ($1 \leq k \leq M - 2m + 2$) distribution, the first group contains the first $(m - 1) + k$ entries and the other group contains the remaining $M - m - k$ entries. For each split, the total area, the sum of edges and the overlap-area of the two new covering rectangles are used to determine the split. Note that not all of three can be minimized at the same time. In [18], three selection criteria were proposed based on the minimum over one dimension, the minimum of the sum of the three values over one dimension or one sort, and the overall minimum. In the algorithm, the minimization of the edges is used.

Dynamic hierarchical spatial indexes are sensitive to the order of the insertion of data. A tree may behave differently for the same data set with a different sequence of insertions. Data rectangles inserted previously may result in a bad split in R-tree after some insertions. Hence it may be worth doing some local reorganization, which is however expensive. The R-tree deletion algorithm provides reorganization of the tree to some extent, by forcing the entries underflowed to be inserted from the root. The study in [18] shows that the deletion and reinsertion can improve the R-tree quite significantly. Using the idea of reinsertion of the R-tree, Beckmann et al. proposed a reinsertion algorithm when a node overflows. The reinsertion sorts the entries in decreasing order of the distance between the centroids of the rectangle and the covering rectangle and reinserts the first p (variable for tuning) entries. In some cases, the entries are reinserted back into the same node and hence a split is eventually necessary. The reinsertion will no doubt increase the storage utilization. But it can be fairly expensive when the tree is large. In the experiments conducted in [18], the R*-tree is found to be more efficient than some other variants, and the R-tree with linear splitting algorithm is substantially less efficient than the one with quadratic splitting algorithm. In general, the R*-tree is an improvement over the R-tree at the expense of expensive insertion.

---

[1] M is the fan-out of R*-tree, m is the minimum number of index entries (data item) contained by one node in R *-tree.

2.2.3 R^+-Tree

The R^+-tree [19] is a compromise between the R-tree and the K-D-B-tree [5] and was proposed to overcome the problem of the overlapping covering rectangles of internal nodes in the R-tree. The R^+-tree structure is exactly the same as that of the R-tree, however the constraints are slightly different.

- Nodes in an R^+-tree are not guaranteed to be at least half filled.
- The entries of any intermediate (internal) node do not overlap(R-tree allows content rectangles to overlap).
- An object identifier may be stored in more than one leaf node(There are no objects stored twice in R-tree).

The duplication of object identifiers leads to the non-overlapping of entries. The subtrees are searched only if the corresponding covering rectangles intersect the query region. The disjoint covering rectangles avoid the multiple search paths of the R-tree for point queries. For the space in Fig. 2.9, only one path is traversed to search for all objects that contain point p_7, whereas for the R-tree, two search paths exist. However, for certain query rectangles, searching the R^+-tree is more expensive than searching the R-tree.

To insert an object, multiple paths may be traversed. At a node, the subtrees of all entries with covering rectangles that intersect with the object bounding rectangle must be traversed. On reaching the leaf nodes, the object identifier will be stored in the leaf nodes, multiple leaf nodes may store the same object identifier. During an insertion, if a leaf node is full and a split is necessary, the split attempts to reduce the identifier duplications. Similar to the K-D-B-tree, the split of a leaf node may propagate upwards to the root of the tree and the split of a non-leaf node may propagate downwards to the leaves. The split of a node involves finding a partitioning hyperplane to divide the original space into two. The selection of a partitioning hyperplane is supposed to be based on the following four criteria: the clustering of entry rectangles, minimal total x- and y-displacement, minimal total space coverage of two new subspaces, and minimal number of rectangle splits. While the first three criteria aim to reduce the work of searches by tightening the coverage, the fourth criterion confines the height expansion of the tree. The fourth criterion can only minimize the number of covering rectangles of the next lower level that must be split as a consequence. It cannot guarantee that the total number of the split rectangles is minimal. Note that all four criteria cannot possibly be satisfied at the same time.

2.2.4 Hilbert R-Tree

Hilbert R-tree [20], an R-tree variant, is an index for multi-dimensional objects like lines, regions, 3-D objects, or high dimensional feature-based parametric objects. It can be thought of as an extension to B^+-tree for multi-dimensional objects.

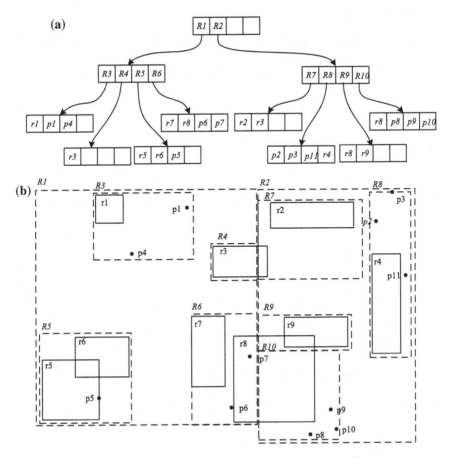

Fig. 2.9 Structure of a R$^+$-tree. **a** Directory of an R$^+$-tree. **b** Structure of a R$^+$-tree

The performance of R-trees depends on the quality of the algorithm that clusters the data rectangles on a node. Hilbert R-trees use space-filling curves, specifically the Hilbert curves, to impose a linear ordering on the data rectangles. There are two types of Hilbert R-trees: one for static databases, and the other one for dynamic databases. In both cases Hilbert space-filling curves are used to achieve better ordering of multi-dimensional objects in the node. This ordering has to be 'good', in the sense that it should group 'similar' data rectangles together, to minimize the area and perimeter of the resulting minimum bounding rectangles (MBRs). Packed Hilbert R-trees are suitable for static databases in which updates are very rare or even no updates at all. The dynamic Hilbert R-tree is suitable for dynamic databases where insertions, deletions, or updates may occur in real time. Moreover, dynamic Hilbert R-trees employ flexible deferred splitting mechanism to increase the space utilization. Every node has a well-defined set of sibling nodes. The Hilbert R-tree sorts rectangles according to the Hilbert value of the center of the rectangles (i.e., MBR). (The Hilbert value of a point is the length of the Hilbert curve from the origin to the point.) Given the ordering, every node has a well-defined set of sibling nodes. Thus, deferred splitting can be used. By adjusting the split policy, the Hilbert R-tree can achieve as high utilization as desired. To the contrary, other R-tree variants have no control over the space utilization.

2.2.5 P-Tree

In many applications, intervals are not a good approximation of the data objects enclosed. In order to combine the flexibility of polygon-shaped containers with the simplicity of the R-tree, Jagadish [21] and Schiwietz [22] independently proposed different variants of polyhedral trees or P-trees. The P-tree of Jagadish uses multi-attribute search structures for polyhedral regions, by mapping polyhedral regions into rectangular regions of a higher dimension. It first introduces a variable number m of orientations in the d-dimensional universe, where $m > d$. Objects are approximated by minimum bounding polytopes whose faces are parallel to these m orientations. We can map the original space into an m-dimensional orientation space, such that each (d-dimensional) approximating polytope P^d turns into an m-dimensional interval I^m. Any point inside (outside) P^d maps onto a point inside (outside) I^m, whereas the opposite is not necessarily true.

The P-tree of Schiwietz (called SP-tree) chooses a slightly different approach to store polygonal objects that tries to combine the advantages of the cell tree and the R*-tree for the two-dimensional case. Basically, the SP-tree is an R-tree whose interior nodes correspond to a nesting of polytopes rather than just rectangles. In general, the number of vertices (and therefore the storage requirements) of a polytope is not bounded. Moreover, when used for approximating other objects, the accuracy of the approximation is positively correlated with the number of vertices of the approximating convex polygon. On the other hand, when used as index entries, there should be an upper bound in order to guarantee a minimum fan-out of the interior nodes.

2.3 Quad-Tree Based Structures

A quad-tree is a tree data structure in which each internal node has exactly four children. Quad-trees are most often used to partition a two-dimensional space by recursively subdividing it into four quadrants or regions. The regions may be square or rectangular, or may have arbitrary shapes. This data structure was named a quad-tree by Raphael Finkel and Bentley in 1974 [23]. There are three typical kinds of quad-trees: point quad-tree, region-based quad-tree (MX quad-tree and PR quad-tree) and CIF quad-tree. Point quad-tree and region-based quad-tree index points in space, while CIF quad-tree is proposed for representing a set of small rectangles for

VLSI (very large scale integration) applications. All forms of quad-trees share some common features:

- They decompose space into adaptable cells.
- Each cell (or bucket) has a maximum capacity. When the maximum capacity is reached, the bucket is split.
- The tree directory follows the spatial decomposition of the quad-tree.

2.3.1 Point Quad-Tree

The point quad-tree [23] is a multi-dimensional generalization of a binary search tree. In two dimensions, each data point is a node in a tree having four sons which are roots of subtrees corresponding to quadrants labeled in order of NE, NW, SW, and SE (shown in Fig. 2.10). Each data point is assumed to be unique. The process of data point quad-trees is analogous to that used for binary search trees. In essence, we search for the desired record on the basis of its x and y coordinates. At each node of the tree, a four-way comparison operation is performed and the appropriate subtree is chosen for the next test. Reaching the bottom of the tree without finding the record means that it should be inserted at this position. The shape of the resulting tree depends on the order that records are inserted. For example, the tree in Fig. 2.10 is the point quad-tree for the sequence of Chicago, Mobile, Toronto, Buffalo, Denver, Omaha, Atlanta, and Miami. Deletion of a node is more complex when the tree is not balanced.

Point quad-trees are especially attractive in applications that involve search. However, they have also been used to solve a measure problem with rectangular ranges in three-dimension. A typical query is that requests the determination of all records within a specified distance of a given record, for example, all cities within 50 miles of Washington, D.C. The efficiency of the point quad-tree lies in its role as a pruning device on the number of searches that is required. Thus many records need not to be examined. For example, supposing that in the hypothetical database of Fig. 2.10, we wish to find all cities within eight units of a data point with coordinates (83, 10). In such a case, there is no need to search the NW, NE, and SW quadrants of the root (i.e., Chicago with coordinates (35, 40)). Thus we can restrict our search to the SE quadrant of the tree rooted at Chicago. Similarly, there is no need to search the NW and SW quadrants of the tree rooted at Mobile (i.e., coordinates (50, 10)).

2.3.2 MX Quad-Tree

Although conceivably there are many ways to adapt the region quad-tree to represent point data, our discussion is limited to two methods. The first method assumes that the domain of data points is discrete, they are treated as if they are BLACK pixels

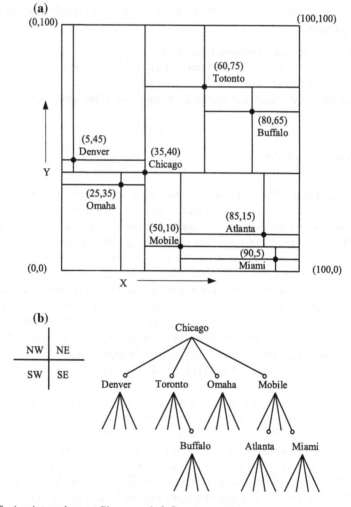

Fig. 2.10 A point quad-tree. **a** Planar graph. **b** Structure graph

in a region quad-tree. An alternative characterization is to think of the data points as nonzero elements in a square matrix. The resulting data structure is called an MX quad-tree (MX for matrix). The MX quad-tree is organized in a similar way to the region quad-tree. The difference is that leaf nodes are BLACK or empty (i.e., WHITE) corresponding to the presence or absence, respectively, of a data point in the appropriate position in the matrix. For example, Fig. 2.11 is the 2^3 by 2^3 MX quad-trees corresponding to the data of Fig. 2.10. It is obtained by applying the mapping f such that $f(Z) = Z$ div 12.5 to both x and y coordinates. The result of the mapping is reflected in the coordinate values in the figure.

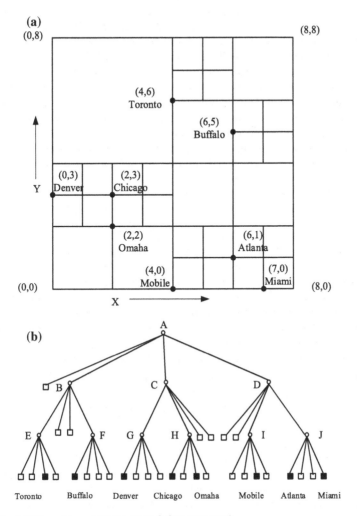

Fig. 2.11 A MX quad-tree. **a** Planar graph. **b** Structure graph

Each data point in an MX quad-tree corresponds to a 1 by 1 square. For ease of notation and operation using modulo and integer division operations, the data point is associated with the lower left corner of the square. This adheres to the general convention followed throughout this presentation that the NE and SE quadrants are closed with respect to the x coordinate and the NW and NE quadrants are closed with respect to the y coordinate. Note that nodes corresponding to data points are not merged, whereas this is not the case for empty leaf nodes. For example, the NW and NE sons of node D in Fig. 2.11 are NIL and likewise for the NW son of nodes corresponding to data points as this results in a loss of the identifying information about the data points. Recall that each data point is different, whereas the empty leaf nodes have the absence of information as their common property and thus can be safely merged. Data points are inserted into an MX quad-tree by searching for them. This search is based on the location of the data point in the matrix (e.g., the discretized values of its x and y coordinate in the example of Fig. 2.11). An unsuccessful search terminates at a leaf node. If this leaf node is NIL, the space spanned by it may have to be repeatedly subdivided until it is a 1 by 1 square. This process is termed splitting and for a 2^n by 2^n MX quad-tree, it will have to be performed at most n times. The shape of the MX quad-tree is independent of the order that data points are inserted. Deletion of nodes is slightly more complex and may require collapsing of nodes–the direct counterpart of the node, that is splitting process outlined above.

2.3.3 PR Quad-Tree

The MX quad-tree is adequate as long as the domain of the data points is discrete and finite. If this is not the case, then the data points cannot be represented since the minimum separation between the data points is unknown. This leads us to an alternative adaptation of the region quadtree to point data that associates data points (that need not be discrete) with quadrants. We call it a PR quad-tree (P for point and R for region). The PR quad-tree is organized in the same way as the region quad-tree. The difference is that leaf nodes are either empty (i.e., WHITE) or contain a data point (i.e., BLACK) and its coordinates. A quadrant contains at most one data point. For example, Fig. 2.12 is the PR quad-tree corresponding to the data of Fig. 2.11. Data points are inserted into PR quad-trees in a manner analogous to that used to insert in a point quad-tree, that is, a search is made for them. Actually, the search is for the quadrant in which the data point, say A, belongs (i.e., a leaf node). If the quadrant is already occupied by other data point with different x and y coordinates, say B, then the quadrant must repeatedly be subdivided (termed splitting) until nodes A and B no longer occupy the same quadrant. This may result in many subdivisions, especially if the Euclidean distance between A and B is very small. The shape of the resulting PR quad-tree is independent of the order that data points are inserted. Deletion of nodes is simple and will not affect other branches, but may require collapsing of nodes, that is, the direct counterpart of the node-splitting process outlined above.

2.3.4 MX-CIF Quad-Tree

The MX-CIF quad-tree is a quad-tree like data structure devised by Kedem [24] (and called a quad-CIF tree, where CIF denotes Caltech Intermediate Form) for representing a large set of very small rectangles for application in VLSI design rule checking. The goal is to locate rapidly a collection of all objects that intersect a given rectangle. The same problem is to insert a rectangle into the data structure under the restriction that it does not intersect existing rectangles. The MX-CIF quad-tree is organized in a similar way to the region quad-tree. A region is repeatedly

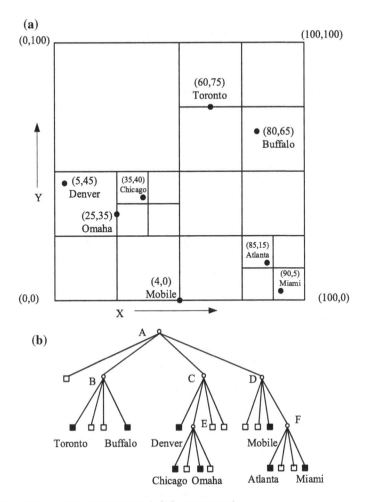

Fig. 2.12 A PR quad-tree. **a** Planar graph. **b** Structure graph

subdivided into four equal-sized quadrants until blocks that do not contain rectangles are obtained. As the subdivision takes place, a set containing all of the rectangles that intersects the lines passing through it is associated with each subdivision point (In other words, rectangles only belong to the minimum corresponding to quadrant which surrounds them). For example, Fig. 2.13 contains a set of rectangles and its corresponding MX-CIF quad-tree. Once a rectangle is associated with a subdivision point, say P, it is not considered to be a member of any son of the node corresponding to P. For example, in Fig. 2.13, node D spans a space that contains both rectangles 11 and 12. However, only rectangle 11 is associated with node D, whereas rectangle 12 is associated with node F.

(a)

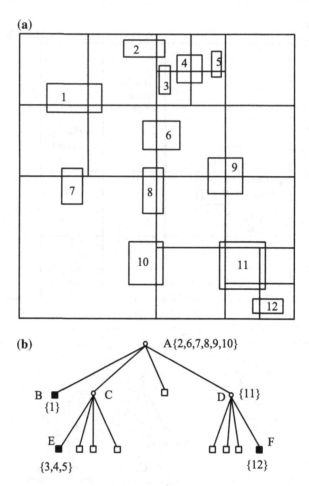

(b)

Fig. 2.13 A MX-CIF quad-tree. **a** Planar graph. **b** Structure graph

The MX-CIF quad-tree is very similar to the MX quad-tree with the following differences. First, data are associated with both terminal and nonterminal nodes. Nevertheless, the analog of a WHITE node is present and is a NIL pointer in the direction of a quadrant that contains no rectangles. Second, we are representing rectangles rather than points. This is fortunate because it provides a termination condition for the subdivision process in forming an MX quad-tree. The nonzero width of the rectangles ensures that they overlap with the subdivision points.

The set of the rectangles that intersects the lines passing through a subdivision point is subdivided into two sets. For example, consider subdivision point P centered at (CX, CY) that partitions a $2 \cdot LX$ by $2 \cdot LY$ rectangular area. All input rectangles that intersect the line $x = CX$ form one set, and all input rectangles that intersect the line $y = CY$ form the other set. Equivalently, these sets correspond to the rectangles intersecting the y and x axes, respectively, passing through (CX, CY). If a rectangle intersects both axes (i.e., it contains the subdivision point P), then we adopt the convention that it is stored with the set associated with the y axis. These subsets are implemented as binary trees, which in actuality are one-dimensional analogs of the MX quad-tree.

Rectangles are inserted into an MX-CIF quad-tree by searching the position that they are to occupy. We assume that the input rectangle does not overlap any existing rectangles. This position is determined by a two-step process. First, the first subdivision point must be located such that at least one of its axis lines (i.e., the quadrant lines emanating from the subdivision point) intersects the input rectangle. Second, having found such a point and an axis, say point P and axis V, the subdivision process is repeated for the V axis until the first subdivision point that is contained within the rectangle is located. During the process of locating the destination position for the input rectangle, the space spanned by the MX-CIF quad-tree may have to be repeatedly subdivided (termed splitting), creating new nodes in the process. As is the case for the MX quad-tree, the shape of the resulting MX-CIF quad-tree is independent of the order that the rectangles are inserted. Deletion of nodes is more complex and may require collapsing of nodes, that is, the direct counterpart of the node-splitting process outlined above.

Compared with MX quad-tree and PR quad-tree, CIF quad-tree can be used to index any shape target without target approximation or space object mapping so that it is more efficient at querying for region. However, region query needs to access the buckets which store points. When the number of indexes increases, as large regions contain more data rectangle, external I/O cost would be expensive.

2.4 Cell Methods Based on Dynamic Hashing

Both extendible hashing [25] and linear hashing [26, 27] lend themselves to an adaptable cell method for organizing k-dimensional objects. The grid file [28–30] and the EXtendible CELL (EXCELL) method [31, 32] are extensions of dynamic hashed organizations [25] incorporating a multi-dimensional file organization for multi-attribute point data.

2.4.1 Grid File

The *gridfile* structure proposed in [28, 30] consists of two basic structures: k linear *scales* and a k-dimensional *directory* (see Fig. 2.14). The fundamental idea is to partition a k-dimensional space according to an orthogonal grid. The grid on a k-dimensional data space is defined as scales which are represented by k one-dimensional arrays. Each boundary in a scale forms a $(k-1)$-dimensional hyperplane that cuts the data space into two subspaces. Boundaries form k-dimensional un-

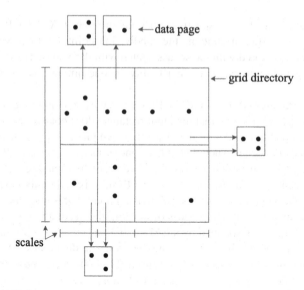

Fig. 2.14 Grid file layout

partitioned rectangular subspaces, which are represented by a k-dimensional arrays known as the grid directory. The correspondence between directory entries and grid cells (blocks) is one-to-one. Each grid cell in the grid directory contains the address of a secondary page and the data page, where the data objects within the grid cell are stored. As the structure does not have the constraint that each grid cell must at least contain m objects, a data page is allowed to store objects from several grid cells as long as the union of these grid cells together from a rectangle, which is known as the storage region. These regions are pairwise disjoint, and together they span the data space. For most applications, the size of the directory dictates that it is stored on secondary storage. However, the scales are much smaller and may be cached in main memory.

Like other tree structures, splitting and merging of data pages are respectively required during insertion and deletion. Insertion of an object entails determining the correct grid cell and fetching the corresponding page followed by a simple insertion of the data page which is not full. When the page is full, a split is required. The split is simple if the storage region covers more than grid cells and the data in the region fall within the same cell. The grid cells are allocated to the existing data page and a new page with the data objects is distributed accordingly. However, if the page region covers only one grid cell or the data of a region fall within only one cell, then the grid has to be extended by a $(k-1)$-dimensional hyperplane that partitions the storage region into two subspaces. A new boundary is inserted into one of the k grid-scales to maintain the one-to-one correspondence between the grid and the grid directory. A $(k-1)$-dimensional cross-section is added into the grid directory. The resulting two storage regions are disjoint and a corresponding data page is attached to each region. The objects stored in the overflowing page are distributed among the two pages: one new page and one existing page. Other grid cells that are partitioned by the new hyperplane are unaffected since both parts of the old grid cell will now share the same data page.

Deletions may cause occupancy of a storage region to fall below an acceptable level, and then trigger merging operations. When the joint occupancy of a storage region whose records have been deleted and its adjacent storage region drops below a certain threshold, the data pages are merged into another one. In [30], based on the average bucket occupancy obtained from the simulation studies, 70 % is suggested to be an appropriate occupancy. Two different methods were proposed for merging, the *neighbor* system and the *buddy* system. The neighbor system allows two data pages whose storage regions are adjacent to merge so long as the new storage region remains rectangular, this may lead to "dead space" where neighboring pages prevent any merging for a particular underpopulated page. A more restrictive merging policy like the buddy system is required to prevent the dead space. For the buddy system, two pages can be merged provided their storage regions can be obtained from the subsequent larger storage region using the splitting process. However, the total elimination of dead space for a k-dimensional space is not always possible.

2.4.2 R-File

As an attempt to improve the performance of grid file, Hutflesz et al. [33] proposed an alternative scheme called the R-file based on the concept of multi-layer files [34]. The R-file is different from the multi-layer grid file in that the R-file has only one layer and is intended for non-zero sized objects. In the R-file, cells are partitioned using the partitioning strategy of the grid file [30] and a cell is split when overflowed. In order for cells to tightly contain the spatial objects, cells are partitioned recursively by repeated halving till the smallest cell that encloses the spatial objects. Spatial objects that are totally contained in a cell are stored in its corresponding data page, and those that intersect the partitioning line are stored in the original cell. If the number of spatial objects that intersect a partitioning line is more than what can be stored in a data page, partitioning line along the other dimensions will be used. If all records lie on the cross point of partitioning lines, they cannot be partitioned by any partitioning lines. In such a case, a chain of buckets is used.

After a split, the original cell and the two new cells overlap, and to keep the directory small, the empty cells are not to maintain. Also, after a split, both the original and new cells have almost the same number of spatial objects. Figure 2.15 illustrates a case in point. Even so, a high number of cells will be inspected for intersection queries, especially in those original large cells. The fact that spatial objects stored in the original un-partitioned cells tend to intersect the partitioning line of the cells suggests the clustering property of these objects. In order to make intersection search more efficient, two extra values that bound the objects in the partitioning dimension are kept within the original cells. Due to the overlapping cells,

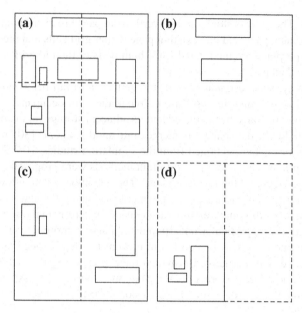

Fig. 2.15 The R-file. **a** Original space. **b** First bucket. **c** Second and third bucket. **d** Fourth bucket

the directory is potentially large. To avoid storing the cell boundaries, a z-ordering scheme [35] is used to number the cells. With such a scheme, cells are partitioned cyclically. For a split not according to the cycle, additional information is stored so that one dimension can be skipped. For each cell, the directory stores the cell number, the bounding interval, and the data bucket reference. The experiments conducted in [33] strongly indicate that the bounding information leads to substantial saving of page accesses.

2.5 Spatial Objects Ordering

Existing DBMS supports efficient one-dimensional indexes and provides fast access to one-dimensional data. If multi-dimensional objects can be converted to one-dimensional objects, such indexes can be used directly without alteration. The functions used in mapping must preserve the proximity between data so as to be well enough to yield reasonably good spatial search. The idea is to assign a number to each of the representative grids in the space and these numbers are then used to obtain a representative number for the spatial objects. In this section, we shall review three different mappings.

2.5.1 Z-Order Curve

Z-order [36], Morton order or Morton code, is a function which maps multi-dimensional data to one dimension while preserving the locality of the data pointer. It was introduced in 1966 by G.M. Morton. The Z-order of a point in multi-dimension is simply calculated by interleaving the binary representations of its coordinate values. Once the data are sorted into this ordering, any one-dimensional data structure can be used such as binary search trees, B-trees, skip lists or hash tables. The resulting ordering can equivalently be described as the order one would get from a depth-first traversal of a quad-tree, the Z-order can be used to efficiently construct quad-trees and related higher dimensional data structures.

As mentioned, the Z-order can be used to efficiently build a quad-tree for a set of points. The basic idea is to sort the input set according to Z-order. Once sorted, the points can be stored in a binary search tree and used directly, which is called a linear quad-tree or they can be used to build a pointer based on quad-tree.

The input points are usually scaled in each dimension to be positive integers, either as a fixed point representation over the unit range [0, 1] or corresponding to the machine word size. Both representations are equivalent and allow for the highest order non-zero bit to be found in constant time. Each square in the quad-tree has a side length which is a power of two, and corner coordinates which are multiples of the side length. Given any two points, the derived square for the two points is the smallest square covering both points. The interleaving of bits from the x and y components of each point is called the shuffle of x and y, and can be extended to higher dimensions.

Points can be sorted according to their shuffle without explicitly interleaving the bits. To do this, for each dimension, the most significant bit of the exclusive or of the coordinates of the two points for that dimension is examined. The dimension for which the most significant bit is largest is then used to compare the two points to determine their shuffle order.

The exclusive operation masks off the higher order bits for which the two coordinates are identical. Since the shuffle interleaves bits from higher order to lower order, the coordinate with the largest significant bit, identifies the first bit in the shuffle order which differs, and that coordinate can be used to compare the two points.

Once the points are in sorted order, two properties make it easy to build a quad-tree: The first is that the points are contained in a square of the quad-tree from a contiguous interval in the sorted order. The second is that if more than one child of a square contains an input point, the square is the derived square for two adjacent points in the sorted order.

For each adjacent pair of points, the derived square is computed and its side length is determined. For each derived square, the interval containing it is bounded by the first larger square to the right and to the left in sorted order. Each interval corresponds to a square in the quad-tree. The result is a compressed quad-tree, where only nodes containing input points or two or more children are present. A non-compressed quad-tree can be built by restoring the missing nodes, if desired.

Rather than building a pointer-based quad-tree, the points can be maintained in sorted order in a data structure such as a binary search tree. This enable points to be added and deleted in $O(\log_2^n)$ time. Two quad-trees can be merged by merging the two sorted sets of points and removing duplicates. Point location can be done by searching for the points preceding and following the query point in the sorted order. If the quad-tree is compressed, the predecessor node found may be an arbitrary leaf inside the compressed node of interest. In this case, it is necessary to find the predecessor of the least common ancestor of the query point and the leaf found.

2.5.2 Hilbert Curve

A Hilbert curve [37] is a continuous fractal space-filling curve first discribed by the German mathematician David Hilbert, as a variant of the space-filling curves discovered by Giuseppe Peano [38].

Because it is space-filling, its Hausdorff dimension is two (precisely, its image is the unit square, whose dimension is two in any definition of dimension; its graph is a compact set homeomorphic to the closed unit interval, with Hausdorff dimension two). Hn is the n-th approximation to the limiting curve. The Euclidean length of Hn is $2^n - 1/2^n$.

Both the true Hilbert curve and its discrete approximations are useful because they give a mapping between 1-D and 2-D space that fairly well preserves locality. If (x, y) are the coordinates of a point within the unit square, and d is the distance along the curve when it reaches that point, then points that have nearby d values will also have nearly (x, y) values. The converse cannot always be true. There will sometimes be points where the (x, y) coordinates are close but their values are far apart. This is inevitable when mapping from a 2-D space to a 1-D space. However, the Hilbert curve does a good job of keeping those d values close together most of the time. So the mappings in both directions do a fairly good job of maintaining locality.

2.6 Summary

Data query and data retrieval is one of the common features of spatial data. Query optimization and solution depend largely on how the data is stored in physical level. Index can accelerate queries and this chapter introduces B-tree and some other typical spatial index such as kd-tree, R-tree, Quad-tree and spatial objects ordering methods and their variants.

Kd-tree is the extension of binary-tree in multi-dimension and it inherits the advantages of binary-tree which performs well in point query. While, the balance of kd-tree is not satisfactory. R-tree is the extension of B-tree in multi-dimension and has good balance. However, the update process of R-tree is complex. The structure of Quad-tree is clear and it is easy to establish the level of the index. As the buckets of

Quad-tree are square, there would be a big difference between depths of each branch. Spatial objects ordering maps multi-dimensional objects to one-dimensional space and uses one-dimensional index to query data. The distance between two objects in one-dimensional space cannot reflect the real situation in multi-dimensional space.

Chapter 3
Road Network Model

Road network can be regarded as a graph which is composed of lines and points. In a national system, cross-city path searches which are often carried out by using urban road network and national highway network, synthetically. For example, to search path from Westlake of Hangzhou to Oriental Pearl of Shanghai, we first need to take advantage of the urban road network looking for the best highway entrance path for Shanghai, and then look for a high-speed road to Shanghai, and lastly find the city's path from highway intersection to Oriental Pearl in the use of urban road network. However, in contrast to dense distributed road network, rural road network and highway network are manifested in a larger range of road space, with lower density, different spatial scales and corresponding traffic rules (such a moving object does not change the direction of movement for a long time). This requires to build a multi-levels road network model to support cross-city search of roads which could switch at different levels of detail or at different levels of road network. Section 3.1 introduces the M^2 map information model which is a hierarchical data structure for managing map objects at different scales, and maintaining each scale respectively. Section 3.2 introduces the multi-levels model which is an integrated representation method for the multi-levels of transportation network. It adopts MOR-tree for managing map objects in multiple levels and uses an integrated method for representing the travel junctions (or traffic constraints), travel cost on road segments and turn corners.

3.1 Map Information Model

3.1.1 L-Model and T-Model

Until now, several multi-levels models have been proposed. Leung et al. [39] proposed a model (denoted as L-model) which is capable of representing space in multi-levels of scales in an integrated way. They discussed the abstraction of space via three

© Springer International Publishing Switzerland 2015
J. Feng and T. Watanabe, *Index and Query Methods in Road Networks*,
Smart Innovation, Systems and Technologies 29,
DOI 10.1007/978-3-319-10789-9_3

interrelated hierarchies: the spatial conceptual hierarchy, the entity hierarchy and the feature hierarchy.

In a spatial context, they call properties of all entities in spatial properties. A subset of the properties defines a spatial concept whose extension is a subset of the entities in space. An example is given in Fig. 3.1: Fig. 3.1a depicts a space with nine entities, and Fig. 3.1b is the conceptual hierarchy for that space. The extension of space is $Ext(Space) = \{CityA, CityB, town1, ra, r1, r2, rv1, rv2, rw\}$, the extensions of other concepts are: $Ext(Transportation) = \{ra, r1, r2\}$, $Ext(Waterbody) = \{rv1, rv2, rw\}$, $Ext(Settlement) = \{CityA, CityB, town1\}$, $Ext(Road) = \{r1, r2\}$, $Ext(Railroad) = \{ra\}$, $Ext(River) = \{rv1, rv2\}$, $Ext(Reservoir) = \{rw\}$, $Ext(City) = \{CityA, CityB\}$, $Ext(Town) = \{town1\}$. The conceptual hierarchy has double semantic meanings. One is the extension enclosure, which expresses the *is-a* relation between any two concepts in the hierarchy from bottom to top. Another is attribute inheritance.

The entity hierarchy is defined based on the natural interpretation of the real world. A bigger entity may contain some smaller entities, or in other words, small entities may be combined to form a composite entity. Taking $CityA$, in Fig. 3.2a, as an entity of the concept 'City' in the conceptual hierarchy, the corresponding entity hierarchy is depicted in Fig. 3.2b. '$CityA$' is composed of a city proper, E, and three counties

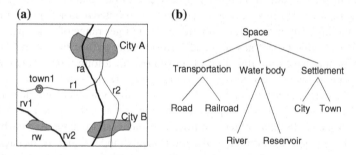

Fig. 3.1 Example of conceptual hierarchy [39]. **a** Map. **b** Hierarchy

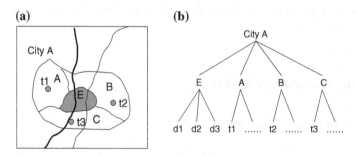

Fig. 3.2 Example of entity hierarchy [39]. **a** Map. **b** Hierarchy

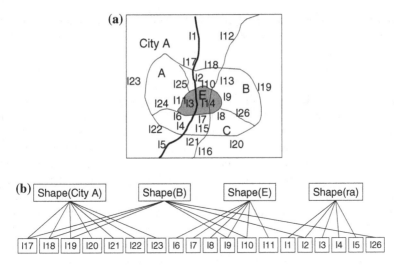

Fig. 3.3 Example of feature hierarchy [39]. **a** Map. **b** Hierarchy

A, *B* and *C*. Furthermore, *E* may be composed of three districts *d*1, *d*2 and *d*3, whereas *A*, *B* and *C* may be composed of towns (*t*1, *t*2, *t*3, etc.).

They express the geometric shapes with two kinds of features, points and lines (an area entity is expressed with its boundary). They decompose shapes of entities into smaller parts and construct feature hierarchy. Figure 3.3 depicts a '*CityA*' with a road *r*1 and a railroad *ra* passing through. The shapes of the entities are formed by the corresponding features as follows: $Shape(CityA) = \{l17, l18, \ldots, l23\}$, and so on.

According to these hierarchies, only the shapes of simple entities at every scale are stored, and the shape of the composite entity can be obtained from the simple entities. The model assures that maps can be displayed customarily. However, every entity possesses spatial information for every scale respectively, and there is no relation among entities in different scales. This model did not support zoom-in/out among multiple scales of maps. In addition, to keep information consistency is complicated when there is modification in a specific area at a specific scale.

To build a multiple presentation database with capabilities for rapid zooming and information consistency, Timpf [40, 41] proposed a hierarchical data structure (denoted as T-model) for managing map objects at different scales, and specified the behavior of map objects over scales. However, the hierarchy they used to arrange spatial entities is too simple to be applied to multiple themes of maps. Their main work was limited to trans-hydro network, which was represented by a filter hierarchy. Therefore, the road segments to be displayed at every scale were based on the same set of road segments at the most detailed scale. In T-model, they distinguished four classes of entities in a map, called map objects: trans-hydro network, containers, areas and map elements.

The trans-hydro network is a merge of the networks of transportation and hydrology. The trans-hydro network creates a partition of map space. Containers are those areas between the lines that the trans-hydro network creates on the map. Areas are a refinement of the container partition and thus need to be contained within a container. Elements are contained in the land-use areas. They introduce a method to describe the structural content of a topographic map series with the help of four different graphs: the trans-hydro graph, the container graph, the area graph, and the element graph. The idea is that every map object is represented in a tree or graph, i.e., they are designed for multiple high level map objects and low level map objects, organized in decreasing levels of detail. This included that an object may be split into sub-objects. Map objects at the same depth of the tree belong to the same map. The trans-hydro graph in Fig. 3.4a shows on the left hand side a schematic representation of the map objects represented in the trans-hydro graph. On the right, each circle represents a segment of a street. The colors signify that the segments belong to the same street. Two segments are filtered from level (d) to level (c), another five segments are filtered from level (c) to level (b). Figure 3.4b shows only a part of a map. Each partition of a map produces at least one tree. The collection of all trees in a map is called the container graph. If the map is considered as the top container, then the container graph will always have the structure of a tree. Figure 3.5c shows an area graph consisting of three trees. Each area is represented by a box in the color and with the symbols of the original area. Lines between boxes signify the relation contains/part-of in the container graph. The area graph has the structure of a generalization hierarchy: land-use classes are generalized. The spatial representation of the areas is aggregated. In Fig. 3.5d, the elements contained in an area are either

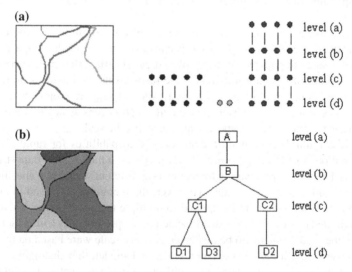

Fig. 3.4 Graphs in T-model (1) [41]. **a** Trans-hydro graph. **b** Container graph

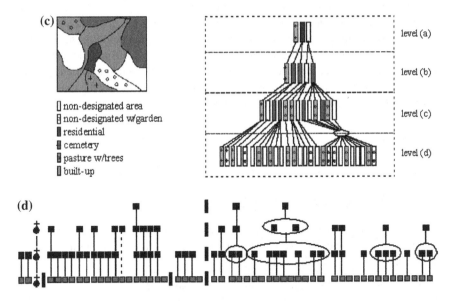

non-designated area
non-designated w/garden
residential
cemetery
pasture w/trees
built-up

level (a)

level (b)

level (c)

level (d)

Fig. 3.5 Graphs in T-model (2) [41]. **c** Area graph. **d** Element graph

buildings, symbols, or dead-ends. The graph created when progressing through the levels of details and linking these elements is called element graph.

Representing country-wide map information in this model results in two problems. One is that only limited themes can be arranged under their model. And another problem is that at the less detailed scale the map is displayed by using the most detailed data. Comparing with the customary map at that scale, there has been too many objects (e.g., nodes which split a road into road segments) displayed. This is a shortcoming for displaying maps on a specific scale and for doing search on the maps. Spaccapietra et al. [42] and Parent et al. [43–45] proposed the MADS conceptual model for managing multiple representations of the same real world entity at different scales. In this model, the spatiality may be associated to object types, attributes, relationships, and aggregation links, and topological relationships between objects are described explicitly. However, there are problems in this model: the redundancy among multi-representations and the complications of modifying topological relations.

Researches of multi-scales maps have also been done from the viewpoint of generating more suitable maps dynamically for a specific purpose. A map synthesis system was proposed by [46], called dynamic information synthesis, because the system helps a user to specify an exact query which corresponds to the required map. They use geographic domain thesaurus which contains aggregation and other semantic relationships as well as conventional thesaurus hierarchy. They define compatibility levels between classes (or concepts) in the geographic domain thesaurus and use it to estimate similarity between logical descriptions. They also introduce Geographic Domain Hierarchical Levels to find and solve the incorrect combinations of words as

ad hoc queries for map generation. All geographic objects have their physical and logical scales, which are defined based on physical attribute (area) and logical attribute
(distribution) of geographic objects. However, the relationship of spatial information
about the geographical objects on different scales has not been studied, which leads to
a result that either the spatial information of a map object is only depicted on a specific
scale or there is redundant spatial information about a map object on several scales.

3.1.2 M^2 Map Information Model

3.1.2.1 Spatial Entity and Map

A *spatial entity* is a term in our model to refer to an object with a shape. The shape
of an entity may be one of or a combination of three basic types in two-dimensional
Euclidean space R^2: node, link and polygon,

(1) Node is a point in two-dimensional Euclidean space R^2.
(2) Link is a straight-line segment whose endpoints are nodes, and no two links
    intersect except for a common endpoint.
(3) Polygon is a union of links, and the interior of a polygon is a closed simply
    connected polygonal region.

A *map* can be defined as a collection of spatial entities, and is drawn according
to a given scale and for a specific region: e.g., a city map on 1:15,000 scale. A *theme
map* is a kind of map which can be defined as a collection of homogeneous spatial
entities: e.g., a road network map in a city of 1:15,000 scale. A *multi-themes map*
can be generated by overlaying several theme maps with common region.

3.1.2.2 Map Region and Associated Hierarchy

Every map has boundary, which limits the map extension to a specific region. Usually,
a map region is defined by the administrative units, and can be decomposed into a
sequence of increasingly finer tessellations.

In our model, a division Σ is defined by a polygon R, called region. A directory
tree is obtained by recursively decomposing a region into a sequence of increasingly
finer tessellations. To define the directory tree formally, a sequence of conditions
$l_0, \ldots, l_h$ is given. A directory tree DT, based on the sequence of $l_0,\ldots,l_h$, is a pair
$(\sum, \leq_d)$, where $\sum = \{\Sigma_0,\ldots,\Sigma_m\}$ is a collection of divisions:

(1) Σ_0 satisfies conditions $l_0,\ldots,l_h$.
(2) The condition l_k such that k = min{q | Σ_i satisfies conditions $l_q,\ldots,l_h$} is called
    the level of division Σ_i, and denoted by $Level(\Sigma_i)$.

(3) For every $1 \leq j \leq m$, Σ_j with region R_j satisfies $Level(\Sigma_j) = l_k$. If there is Σ_i, whose region R_i covers R_j, Σ_i is an ancestor of Σ_j ($\Sigma_i \in Ancestors(\Sigma_j)$), and Σ_j is a descendant of Σ_i ($\Sigma_j \in Descendants(\Sigma_i)$). This relationship is denoted by $\Sigma_j \leq_d \Sigma_i$. If Σ_i satisfies $Level(\Sigma_i) = l_{k-1}$ and R_i covers R_j, Σ_i is the father of Σ_j, and Σ_j is a son of Σ_i. This relationship is denoted by $\Sigma_j = Son(\Sigma_i)$ or $\Sigma_i = Father(\Sigma_j)$.

(4) $\leq_d$ is a partial order on $\sum$ (e.g., reflexive, antisymmetric and transive). For every Σ_i, $l_0 < level(\Sigma_i) \leq l_h$ and $\Sigma_i \leq_d Ancestors(\Sigma_i)$. For every Σ_i, $l_0 \leq level(\Sigma_i) < l_h$ and $Descendants(\Sigma_i) \leq_d \Sigma_i$.

The sequence of conditions $l_0, \ldots, l_h$ can be corresponded to the levels of administrative units: e.g., l_0 for country level and l_1 for prefecture level.

We give an example in Fig. 3.6. Figure 3.6a shows a map series of Japan, Aichi prefecture, Nagoya city and so on with only map boundaries, and Fig. 3.6b gives a

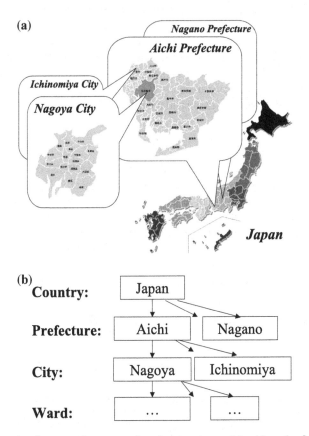

Fig. 3.6 Example of maps and corresponding directory tree. **a** Map hierarchy. **b** Conceptual hierarchy in map

hierarchy of the map series, which is exactly an example of directory tree for the country-wide GIS. Here,

(1) $\sum = \{Japan, Aichi, Nagano, Nagoya, Ichinomiya,...\}$, $\Sigma_0 = Japan$, $\Sigma_1 = Aichi$, $\Sigma_2 = Nagano$, $\Sigma_3 = Nagoya$, $\Sigma_4 = Ichinomiya$.
(2) l_0 corresponds to **Country** level, l_1 corresponds to **Prefecture** level, and so on. $level(\Sigma_0) = level(Japan) = l_0$, $level(\Sigma_1) = level(Aichi) = l_1$, $level(\Sigma_2) = level(Nagono) = l_1$, $level(\Sigma_3) = level(Nagoya) = l_2$, and so on.
(3) $\leq_d$ corresponds to a $sub - region$ of relation between levels, where $Aichi \leq_d Japan$, $Nagano \leq_d Japan$, and $Nagoya \leq_d Aichi$, and so on.

3.1.2.3 Map Theme and Theme Hierarchy

There are many themes in a country-wide GIS, and even inside one theme there are many different detailed maps. For example, a national organization manages the highways all over the country on a country map scale, and a prefectural organization manages more detailed road information including the highway information and prefectural road information inside the prefecture. There is also a hierarchy consistent to the directory tree. We formalize this kind of hierarchy as a theme tree. In a country-wide GIS there are many theme trees.

A section ε is defined as a set of spatial entities $SE = \{...se_k,...\}$. Co is a function defined as:

$$Co : \Sigma_i \rightarrow \{..., \varepsilon_j, ...\} \tag{3.1}$$

It means that there may be many sections (on theme trees) corresponding to a division (on directory tree) of a specific region at a specific level. An example is given in Fig. 3.7. The highways in Japan are defined as a set of highways, which are located inside the region of country "Japan" and are managed by an organization at the country level. So, the section of Japan highways corresponds to the country level of division "Japan"; and the prefecture road in Aichi prefecture corresponds to the prefecture level of division "Aichi".

The theme tree TT is a triple $(\epsilon, \leq_t, f)$, where

(1) ϵ is a finite set of sections $\{\varepsilon_0, ..., \varepsilon_m\}$.
(2) $\leq_t$ is an order on ϵ defined as: $\forall \varepsilon_i, \varepsilon_j \in \epsilon$, $\exists \Sigma_i$, $\Sigma_j \in \sum$, $\varepsilon_i \in Co(\Sigma_i)$, $\varepsilon_j \in Co(\Sigma_j)$,

$$\Sigma_j \leq_d \Sigma_i \Leftrightarrow \varepsilon_j \leq_t \varepsilon_i \tag{3.2}$$

and there are corresponding definitions of Descendants and Ancestors. We define $Region(\varepsilon_i) \equiv R_i$ and $Level(\varepsilon_i) \equiv Level(\Sigma_i)$.
(3) f is a set of functions defined as: $\varepsilon_i, \varepsilon_j \in \epsilon$, $\varepsilon_j \leq_d \varepsilon_i$,

$$f(\varepsilon_i, \varepsilon_j) \rightarrow \varepsilon_{ij} \tag{3.3}$$

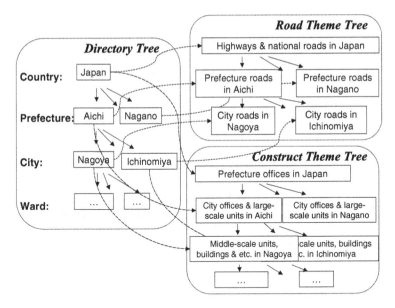

Fig. 3.7 Example of directory tree and theme trees

f works on two sections inside a theme tree and generates a set of spatial entities of ε_{ij}, which is a union of spatial entities in ε_i and ε_j based on predefined relations. All the spatial entities of ε_{ij} are represented in $Region(\varepsilon_j)$ and at the same scale of $Level(\varepsilon_j)$.

There are two kinds of relations defined on the spatial entity subsets of sections which belong to different levels:

(1) *Supplement*: the spatial entity subset of lower level is a supplement of the subset of upper level. In other words, the objects in the upper level are needed by the lower level, while objects in the lower level are not needed by the upper level when generating theme map.

(2) *Generalization*: the subset of upper level can be derived from the subset of the lower level by merging some objects so that some distinct objects in lower level are regarded as indistinguishable objects and become just one object in the upper level.

For the example given in Fig. 3.7, the *Directory Tree* is the same as that in Fig. 3.6, where $\Sigma_0 = Japan$, $\Sigma_1 = Aichi$, and so on. There are sections in *Road Theme Tree* and *Construct Theme Tree*, and

(1) $\varepsilon_1 = Highwaya$ & *national roads in Japan*, $\varepsilon_2 = Prefecture$ *offices in Japan*, $\varepsilon_3 = Prefecture$ *roads in Aichi*, $\varepsilon_4 = Prefecture$ *roads in Nagano*, $\varepsilon_5 = City$ *offices &largescale units in Aichi*, and so on.

(2) The relations among *Directory Tree* and theme trees are: $Co : Japan-> \{\varepsilon_1, \varepsilon_2\}$, $Co : Aichi-> \{\varepsilon_3, \varepsilon_5\}$, and so on.

(3) $\varepsilon_3 \leq_t \varepsilon_1, \varepsilon_4 \leq_t \varepsilon_1, \varepsilon_5 \leq_t \varepsilon_2$ and so on.

(4) In Fig. 3.7, the solid pointer in theme trees corresponds to $\leq_t$ relation, that in *Directory Tree* corresponds to $\leq_d$ relation; the dotted pointer from node of *Directory Tree* to node of theme tree corresponds to the input and output of *Co* function.

3.1.2.4 Formulation of M^2 Model

To represent and manage map information for a country-wide integrated GIS, we address a Multi-scales/Multi-themes (M^2) map information model, which does not only meet the need of integration and retrieval for GIS but also possesses the extensibility and scalability. In M^2 model, map elements are looked upon as objects (the same as spatial entities in this section) and are uniquely assigned to the appropriate theme, scale and region without being duplicated among them.

Informally, the model composes a directory tree, which defines the map regions with corresponding scales, and some theme trees, which define a series of specific theme maps hierarchically. We give a definition of our model in UML (as Fig. 3.8). The structure of the model mainly consists of two kinds of separated and related hierarchies: the directory tree and the theme tree, which can be expressed by a triple

$$M^2 = (D, T, Co) \tag{3.4}$$

Here, D is a set of directory trees (DT), T is a set of theme trees (TT), and *Co* is a mapping from D to T.

M^2 Map Conceptual Model in UML

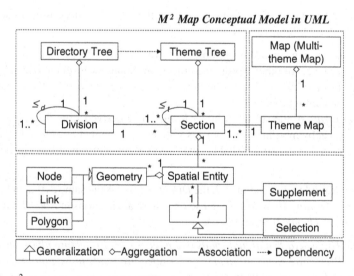

Fig. 3.8 M^2 map conceptual model in UML

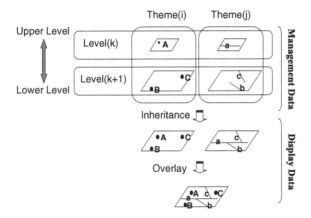

Fig. 3.9 Aspect on Multi-scales/Multi-themes for point and line objects

The model is powerful to integrate various scales of maps uniformly. However, it is necessary to aggregate and arrange different scales of data derived from different levels adaptively and then propagate the information from upper levels to lower level. For the purpose of propagating the objects in upper levels to the lower level, inheritance functions are developed. With the input of upper levels' objects on a small scale and relations between upper and lower levels, inheritance functions produce lower level's theme map on a large scale, and at the same time overlay functions are developed to merge the map elements in different themes into one multi-themes map. That is to say, for a particular application, map information on needed scale can be prepared dynamically by using inheritance functions and overlay functions.

Figure 3.9 shows the aspect of Multi-scales/Multi-themes for point and line objects. In each theme (theme (i) and theme (j)), inheritance functions propagate management data in the upper level (level k) to the lower level (level (k + 1)) and get the display data of level (k + 1). Then, overlay functions merge multi-themes and get a multi-theme map on the scale of level (k + 1).

We define theme map as TM_ε, which is a map of a theme on the section ε, and is denoted as a triple $(SE_{TM}, R_{TM}, L_{TM})$: $SE_{TM} = \{se | se \in SE_\varepsilon \cup SE_{Inherit(\varepsilon)}\}$, $R_{TM} = Region(\varepsilon)$, $L_{TM} = Level(\varepsilon)$.

To generate a theme map for a required region at a specific scale consists of two steps: one is a filtering step, which records division series $\Sigma_0, \ldots, \Sigma_i$ on the directory tree. The regions of these divisions cover the required region. Another is a generation step, which generates the theme map by using inheritance functions, based on the corresponding sections in the theme tree. The inheritance function, $Inherit(\varepsilon_i)$, uses the function f defined in the previous section to transfer all the spatial entities in ancestors of ε_i from the root level $Level(\varepsilon_0)$ to $Level(\varepsilon_i)$, recursively.

The inheritance function is defined as:
==============================
Function *Inherit(ε:* section): section;
Begin
   if $\varepsilon == \varepsilon_0$
     then return ε
     else return $f(Inherit(Parent(\varepsilon)), \varepsilon)$;
end.
==============================

Multi-themes map (MTM) ($SE_{MTM}, R_{MTM}, L_{MTM}$) can be generated by over-laying several themes maps (as Fig. 3.9).

$$MTM = Overlay(TM_{\epsilon_i}, ..., TM_{\epsilon_j}) \tag{3.5}$$

$$SE_{MTM} = \{se | se \in SE_{\varepsilon_i} \cup ... \cup SE_{\varepsilon_j}\} \tag{3.6}$$

$$R_{MTM} = Region(\varepsilon_i) \cap ... \cap Region(\varepsilon_j) \tag{3.7}$$

$$L_{MTM} = min(Level(\varepsilon_i), ..., Level(\varepsilon_j)) \tag{3.8}$$

Data maintenance is as important as data generation. We keep the data consistent among the distributed datasets by predefining relations; i.e., *Supplement* and *Generalization*, among spatial entities of these datasets.

3.1.2.5 Road Representation

Road network is an important theme in a country-wide GIS. The representation of road network under M^2 model leads to no-redundancy management of road network on several scales and supports efficient spatial query on them.

Relations among Objects in Road Theme:
There are two kinds of relations among road objects, belonging to two different levels of theme maps: selection and generalization. Selection means that for an object in the upper level of theme map, there is a corresponding object in the lower level of theme map and generalization is the same as *Generalization* defined in the previous section (Fig. 3.10).

Since in M^2 map information model the information of a spatial entity is stored only once theoretically, road objects are divided into appropriate sections without redundancy. There are four kinds of relations defined among the objects, belonging to different sections (see Fig. 3.11). The first three kinds ((a), (b) and (c)) are *supplement* relations, which are defined based on the *selection* relations between corresponding theme maps, and the relation defined in (d) is *generalization* based on the corresponding relation between corresponding theme maps.

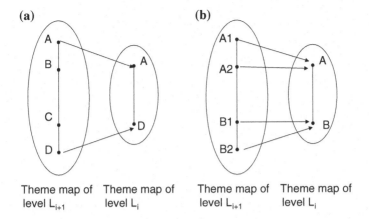

Fig. 3.10 Relations among road objects belonged to two road theme maps. **a** Selection.
b Generalization

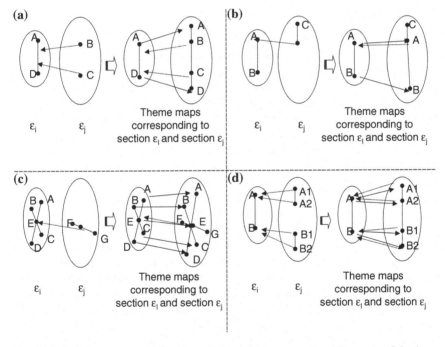

Fig. 3.11 Relations among road objects belonged to two road sections. **a–c** Selection.
d Generalization

Road Representation and Inheritance Mechanism:

In the M^2 model, road information is managed in several levels with different
scales according to the road classes in the road theme: i.e., highways are stored in

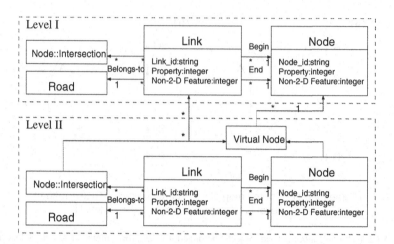

Fig. 3.12 Schema diagram for road-theme database

the country level on a scale of 1:20,000,000; prefecture roads are in the prefecture level on 1:200,000; and city roads are in the city level on 1:15,000, respectively.

Though the scales for levels are different, roads are represented by nodes and links in every level. Figure 3.12 shows the schema diagram of road theme in two levels. There are two kinds of nodes: real node and virtual node. A real node is a spatial entity in M^2 model with point geometry and non-2-D feature. This describes a point in the road network where traffic conditions change. The point can be a crossroad, a traffic circle, a toll, a dead end, an intersection with boundary line, or a point where some values of road attributes change. A node can delimit several road segments and each road segment has only one begin-node and only one end-node. A virtual node is an object in the lower level without any of attribute values, but points to the corresponding real node in the upper level. To access a virtual node is actually to access a real node in the upper level. The definition of real and virtual nodes reduces the representation redundancy in the model. Table 3.1 lists out the main classes of node properties concerning the process among levels. The node with "intersection with upper level" property is a kind of node managed in the lower level, but extends the road of upper level. We can see its effect in the example of road network with two levels, given in Fig. 3.13. The links in the upper level are represented by bold lines. In Fig. 3.13, a real intersection of upper level and lower level is represented as a dotted circle with heart point (e.g., N22 and N29), and an intersection between two levels but the two points are on different 2-D spaces is represented as a dotted circle without heart point (e.g., N23 and N28). The node N26 in the lower level is a virtual node and the corresponding real node is N14 in the upper level.

The main attributes of nodes in level 1 and level 2 are given in Tables 3.2 and 3.3. In this example, the value of non-2-D property means over-pass (1), regular (0) and underground (-1).

Table 3.1 Properties of nodes

Class	Property	Meaning
0	Heritage from upper level	Virtual node; Upper level's end or intersection node There is at least one link connected to it in lower level
1	End	Degree of node is 1 or 2 in lower level
2	Intersection	Degree of node is more than 2 in lower level
3	Intersection with upper level	Node points link of upper level, if they are on the same 2-D space The link is split when inheritance function works
4	$(1+3)$	End-node which is intersection with upper level at the same time
5	$(2+3)$	Intersection which is intersection with upper level at the same time
6–9	Property change tag	Other properties except 0–5 property-change points

Also, there are two kinds of links: regular link and inherited link. A regular link is an atomic object of M^2 model with linear geometry and non-2-D feature: for example it is a road fragment. An inherited link is a line between one level's or two different levels' nodes when there is a link produced by the inheritance function. This is not persistently stored in the database of road network, but only is produced when needed. The main attributes of links in level 1 are given in Table 3.4 and the inherited links in level 2 are given in Table 3.5. In Table 3.5, the inheritance functions of road theme propagate both links and nodes of the upper level to the lower level with regard to the properties and non-2-D features of nodes and links. The nodes N22 and N29 in the lower level are the real intersections between the upper level and the lower level, and split into L11 and L15; the nodes N23 and N26 do not split L14 because they are not in the same 2-D space as L14, the node N20 splits L13(N14, N12) into two links (N14, N20) and (N20, N12) when the inheritance function works and the other links are inherited directly.

In the hierarchical road model the road information is managed in several levels with different scales according to the road classes, and relations among layers are managed in the lower layer. Thus, there are advantages:

(1) It is easy to extend the system. The road hierarchy is implemented first from the top level with the smallest scale, next the lower levels with larger scales and then the relations (pointers) associated with upper levels can be implemented. There are no scale limits in the system.

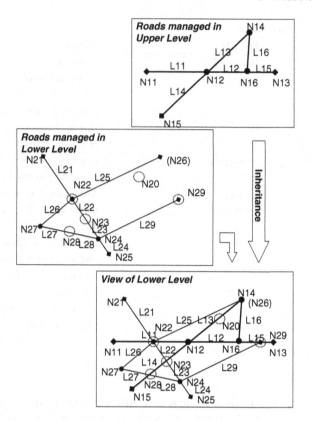

Fig. 3.13 Example of road network with two levels

Table 3.2 Nodes managed in level 1

Node-id	Geometry	Property	Non-2-D feature	Pointer
N11	Point	1	0	
N12	Point	2	0	
N13	Point	1	0	
N14	Point	2	0	
N15	Point	1	0	
N16	Point	2	0	

(2) It is easy to get needed scale view of the map. One can easily access map elements up to the top of the hierarchical level by using inheritance functions.

(3) It is possible to manage road network in multiple scales by using smaller space. At the same time, it is promised to enforce consistency between scales and reduce the global update load.

Table 3.3 Nodes managed in level 2

Node-id	Geometry	Property	Non-2-D feature	Pointer
N21	Point	1	0	
N22	Point	5	0	L11
N23	Point	3	0	L14
N24	Point	2	0	
N25	Point	1	0	
N26	Point	0	0	N14
N27	Point	2	0	
N28	Point	3	0	L14
N29	Point	4	0	L15
N20	Point	3	0	L13

Table 3.4 Links managed in level 1

Link-id	Geometry	Non-2-D feature	Begin-node	End-node
L11	Line	0	N11	N12
L12	Line	0	N12	N16
L13	Line	0	N14	N12
L14	Line	1	N12	N15
L15	Line	0	N16	N13
L16	Line	0	N14	N16

Table 3.5 Inherited link in level 2

Link-id	Geometry	Non-2-D feature	Begin-node	End-node
L11	Line	0	N11	N22
	Line	0	N22	N12
L12	Line	0	N12	N16
L13	Line	0	N14	N20
	Line	0	N20	N12
L14	Line	1	N12	N15
L15	Line	0	N16	N29
	Line	0	N29	N13
L16	Line	0	N14	N16

Topological Search:

The capability of computing path queries is essential in the database manipulation for many advanced applications such as navigation systems, Geographical Information Systems (GIS) [47]. The key of computing path queries is to find out the topological relations among roads. Topological relations are those that are invariant to topological transformations, i.e., relations which are not changed after transformations like rotation, translation, scaling and rubber sheeting [48].

In M^2, the topological relations between map elements through the inheritance and overlay operations may be different from the relations before the operations: e.g., in Fig. 3.9, the place "A" is "joint" with road "a" in the upper level of small scale and is changed to "disjoint" with it in the lower level of large scale. The method of associating topological relations directly with the defined geometry can compute out the spatial relations among map components on the same scales, but cannot process the map components on different scales. On the other hand, the method of looking upon spatial relations as attributes, proposed by [39, 42, 44], would bring complicated relations between levels and themes, and break the easy extensibility of the model. As a result the topological relations between map elements in M^2 model should be computed dynamically.

3.1.2.6 Evaluation

We compare the model (M^2) with the models proposed by Leung et al. [39](L-model) and Timpf [41] (T-model).

The experiment environment is developed in Java (for ensuring our system's portability over different platforms) on SGI O2 R5000 SC 180 entry-level desktop workstation. The system manages themes of country, prefecture and city levels, based on the maps of Japan, Aichi Prefecture (Map 2500 of Geographical Survey Institute [49]).

Highways and national roads are stored in the country level (the first upper level), prefecture roads and main local roads in the prefecture level (the second level), and city roads in the city level (the third level). The node stored in the upper level may be referred by the lower level through the pointer of a virtual node in the lower level, and the node in the lower level splits the road segment of the upper level when using the inheritance functions to propagate the road information on a small scale of upper level to the lower level on a large scale.

Evaluation of Road Network Representation and Path Search:

In order to evaluate the model in representing road network, the quantity of storage links is measured and is displayed links on the road network. Here, the model is compared with L-model and T-model.

(1) *Comparison of M^2 model and L-model.* In the prototype system, the road network is divided into 3 levels without redundancy with regard to the M^2 model. In L-model, the road network is managed on different levels with redundancy with regard to a customary law. Therefore, the two methods provide the same display maps in every level. However, the number of storage links in M^2 model is less than that in L-model on every level.

(2) *Comparison of M^2 model and T-model.* In T-model, the road network is managed only on the most detailed level. The display map of every level is a part of the detailed map. We can observe that there are more road objects displayed in the

upper levels in T-model than those in our model. In M^2 model, we generate display roads on lower levels by using inheritance functions, and display them customarily; while in T-model, because the maps on every level are based on the same dataset of the most detailed map, there are more road segments on the upper levels than those in M^2 model. Generating maps with more objects than needed can result in an uncustomary displaying of maps and inefficient search process on the road network.

In general, M^2 Map Information model outperforms L-model and T-model in supporting the effective storage, display of road network, and efficient spatial query and data maintenance for road network.

3.2 Multi-levels Model for Transportation Network

The management facility/environment of geographic database is one of the important subjects concerned with ITS. A geographic database may contain static map data (including data of road network and other map objects), public transportation routes, and current travel cost (e.g., travel time) on segments of transportation network [50].

In the country-wide system, processing the map and transportation information in many levels of details will be needed. For example, a query of path search between two city halls which belong to two separated prefectures is solved more efficiently by dividing into three sub-queries: one sub-query for finding path between two cities based on the country-level and prefecture-level roads (i.e., highways and crossing-prefecture roads), and two sub-queries based on city-level roads (all the streets inside the city) for finding the path between the city halls and the upper level roads. If this search is required to compute the path length between two places, it can be solved based on the road network datasets generated in M^2 model. However, when the search is required to compute the travel cost based on the current traffic situation, it should be done based on a transportation network, which is different from road maps.

In this section an integrated representation method is proposed for the multi-levels of transportation network. Our method adopts MOR-tree for managing map objects in multiple levels and uses an integrated method for representing the travel junctions (or traffic constraints), travel cost on road segments and turn corners. Based on the datasets created by this method, queries in ITS applications can be processed efficiently by using proper search methods.

3.2.1 Representation of Transportation Information

Transportation information is different from the information of road network. It is important to identify one-way roads with attributes of links, traffic constraints, infor-

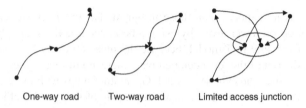

One-way road   Two-way road   Limited access junction

Fig. 3.14 Different types of roads and junctions [53]

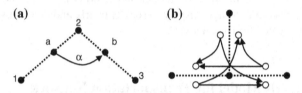

Fig. 3.15 Pseudo-dual graph. **a** A primary graph (*dotted lines*) and its pseudo-dual graph (*real line*), and **b** a T-road primary graph (*dotted lines*) and its restricted pseudo-dual graph (*real line*) [52]

mation about turns between links, or access conditions from one link to another [51]. Moreover, for some important route planning problems, the turn costs are also taken into consideration [52], encountered when we make a turn on a cross-point. A typical method [53] represents the transportation network using a directed graph. In the graph, each edge depicts a one-way road and each node corresponds to a junction. Two-ways roads can be presented as a pair of edges: one in each direction. However, extra nodes should be added to the graph when there are any access limitations (constraints of specific traffic controls). In other words, one node on the road network may be represented with several vertices corresponding to the junctions, and they are independent with each other. Figure 3.14 gives the representation of different types of roads and junctions: one-way road, two-ways road without any access limitations, and T-junction with some access limitations (the center point on this T-junction is represented by two vertices in this directed graph).

Since this representation method ignores the spatial attributes of map objects, only the routing queries are applicable well on this model. For example, Lee's algorithm [54] is used for routing on this model. Lee's algorithm finds the best route with respect to optimizing some metric, as long as a route exists. The method simulates an ink-blot spreading out over a piece of paper. The covered area represents the already explored vertices. When the destination is reached, the algorithm traces the route back to the start.

An architecture was proposed in [55] for keeping traffic information on nodes of road network. However, the information of traffic constraints and turn costs on the nodes is omitted in their discussion. To represent the traffic cost and the turn cost, a method in [52] was proposed. The turn cost is represented by a pseudo-dual graph. Figure 3.15 gives the pseudo-dual graph for representation of turn costs on a primary graph (Fig. 3.15a) and a T-junction with some access limitations (Fig. 3.15b). The turn cost is represented with the additional nodes and links in the pseudo-dual graph.

In (a), two nodes a and b are added to the primary graph, and an arc α from a to b represents the turn cost of leaving the traffic arc from node 1 to node 2 entering the traffic arc from node 2 to node 3; In (b), there are six nodes and six arcs for additionally representing turn costs between possible pairs of traffic arcs. The cost of search algorithms (e.g., Dijkstra's algorithm [56]) becomes high. Moreover, the pseudo-dual graph is insufficient (and needs the reference to the primary graph) for route drawing.

3.2.2 Modeling of Road Network and Traffic Information

Not only the kinds of information but also the management methods of transportation information affect the processing efficiency of queries in ITS applications. In this section, we propose a representation method for integrating traffic information and spatial information about road network by considering the following terms:

(1) The traffic conditions change continuously, and the snapshot of conditions is recorded as traffic information. In comparison with the traffic information, the map of road network is seldom updated, and can be regarded as static information. Therefore, if the static information is managed by an efficient structure, the changes of traffic information associated with the road map should not disturb the stability of the structure.

(2) The integrated representation should support not only the spatial query on road network and the temporal query on traffic information, but also the interaction between these two kinds of queries.

A road network with nodes and links representing respectively the crosses and road segments can be regarded as an un-directed graph G, $G = (V, L)$, where V is a set of vertices $\{v_1, v_2, \ldots v_n\}$, and L is a collection of lines $\{l_1, l_2, \ldots l_m\}$. Traffic information on the road network is regarded as a directed graph G', $G' = (V, A)$, where V is a set of vertices $\{v_1, v_2, \ldots v_n\}$, and A is a collection of arcs $\{a_1, a_2, \ldots a_p\}$.

Figure 3.16 depicts these two kinds of graphs. In the un-directed graph of Fig. 3.16a road segments are represented by lines, while in the directed graph of Fig. 3.16b junctions are represented by arcs. One line for road segment in Fig. 3.16a may be corresponded to two arcs in Fig. 3.16b with two-directions traffic information.

Fig. 3.16 Road segment and traffic arc. **a** Road segment by one link. **b** Road segment by two arcs

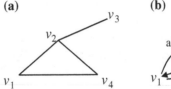 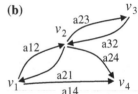

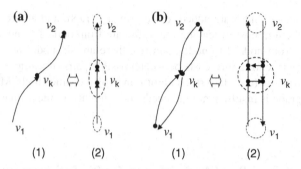

Fig. 3.17 One-way road and two-ways road. **a** One-link road segment. **1** Representation by [53]'s model. **2** Super-node representation. **b** Two-arcs road segment. **1** Representation by [53]'s model. **2** Super-node representation

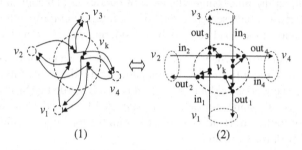

Fig. 3.18 Cross node with constraint. **1** Representation by [53]'s model. **2** Super-node representation

In addition to the directions of traffic, there are usually traffic controls (constraints) on road network to constrain the action of traffic. For example, the right-turn and U-turn are forbidden on some cross-points. The typical road junctions with and without constraints are given in Figs. 3.17 and 3.18. Road junctions are represented by using [53]'s model in Figs. 3.17a(1), b(1) and 3.18(1). Considering the shortcomings of this simple model, we propose *super-node* representation method for integrating junctions (including traffic cost and traffic constraints) and road network.

A *super-node* can be defined as a node in road network with multiple corresponding junctions, for example, v_k in Figs. 3.17a(2), b(2) and 3.18(2). The information on the *super-node* contains the following parts (for simplicity of explanation, road junctions in Fig. 3.18(2) are used as an example):

(1) *Cost-arc*. The arcs which have v_k as their final vertex are called in-arcs, denoted as in_i, and similarly the arcs which have v_k as their initial vertex are called out-arcs, denoted as out_j. The number of those arcs is called in-degree (e.g., 4) and out-degree (e.g., 4), respectively. Every out_i is defined as a *Cost-arc*, which consists of the final vertex of this arc and the traffic cost of this arc. *Cost-arc*s of v_k in Fig. 3.18(2) are

$$\begin{bmatrix} out_1(v_1, cost_{k1}) \\ out_2(v_2, cost_{k2}) \\ out_3(v_3, cost_{k3}) \\ out_4(v_4, cost_{k4}) \end{bmatrix}. \tag{3.9}$$

(2) *Constraint-matrix*. The constraints on the *super-node* can be represented with an $n \times m$ matrix CM:

$$CM(v_k) = \begin{array}{c} \\ in_1 \\ in_2 \\ \vdots \\ in_n \end{array} \begin{array}{cccc} out_1 & out_2 & \dots & out_m \\ \left(\begin{array}{cccc} C_{11} & C_{12} & \dots & C_{1m} \\ C_{21} & C_{22} & \dots & C_{2m} \\ \vdots & \vdots & \ddots & \vdots \\ C_{n1} & C_{n2} & \dots & C_{nm} \end{array} \right) \end{array} \tag{3.10}$$

and

$$C_{ij} = \begin{cases} 1 & there\,is\,restriction\,from\,in_i\,to\,out_j; \\ 0 & there\,is\,a\,junction\,from\,in_i\,to\,out_j. \end{cases} \tag{3.11}$$

Constraint-matrix for v_k in Fig. 3.18(2) is:

$$CM(v_k) = \begin{array}{c} \\ in_1 \\ in_2 \\ in_3 \\ in_4 \end{array} \begin{array}{cccc} out_1 & out_2 & out_3 & out_4 \\ \left(\begin{array}{cccc} 1 & 0 & 0 & 1 \\ 1 & 1 & 0 & 0 \\ 0 & 1 & 1 & 0 \\ 0 & 0 & 1 & 1 \end{array} \right) \end{array} \tag{3.12}$$

where there are restrictions on going from in_1 to out_1 and out_4, from in_2 to out_1 and out_2, from in_3 to out_2 and out_3, and from in_4 to out_3 and out_4. If there is no restriction for any in_i of the super-node v_k, *Constraint-matrix* of v_k is filled with 0, and is regarded as $\emptyset$.

Moreover, our method is able to process the turn cost by extending *Constraint-matrix* to a *Turn-Cost/Constraint-matrix*. *CM* can be modified to a *Turn-Cost/Constraint-matrix*:

$$T\_CM(v_k) = \begin{array}{c} \\ in_1 \\ in_2 \\ \vdots \\ in_n \end{array} \begin{array}{cccc} out_1 & out_2 & \dots & out_m \\ \left(\begin{array}{cccc} TC_{11} & TC_{12} & \dots & TC_{1m} \\ TC_{21} & TC_{22} & \dots & TC_{2m} \\ \vdots & \vdots & \ddots & \vdots \\ TC_{n1} & TC_{n2} & \dots & TC_{nm} \end{array} \right) \end{array} \tag{3.13}$$

and

$$\begin{cases} 0 \le TC_{ij} < Max & the\,turn\,cost\,from\,in_i\,to\,out_j; \\ TC_{ij} = Max & there\,is\,restriction\,from\,in_i\,to\,out_j. \end{cases} \tag{3.14}$$

For example, the *Turn-Cost/Constraint-matrix* for v_k in Fig. 3.18(2) may be like this:

$$T\_CM(v_k) = \begin{array}{c} \\ in_1 \\ in_2 \\ in_3 \\ in_4 \end{array} \begin{array}{cccc} out_1 & out_2 & out_3 & out_4 \\ \left(\begin{array}{cccc} MAX & 10 & 40 & MAX \\ MAX & MAX & 10 & 30 \\ 10 & MAX & MAX & 30 \\ 10 & 40 & MAX & MAX \end{array} \right) \end{array} \qquad (3.15)$$

where MAX is defined as a large constant value. The element TC_{ij} in this matrix with a value MAX represents a restriction from in_i to out_j, e.g., U-turn and right-turn are forbidden in this example. So, MAX is assigned to TC_{ii} ($i = 1, 2, 3, 4$), TC_{14}, TC_{21}, TC_{32} and TC_{43}. The value TC_{12} represents the cost 10 (e.g., 10 s) of making a left-turn on the cross-point (from in_1 to out_2), while the cost of crossing the point v_k from in_1 to out_3 is 40.

This method decreases the redundancies in the database by adopting a complex node representation. It is easy to integrate the traffic information and the basic road network. For the basic road network, the additional information for traffic information is managed on every node. When the number of nodes and traffic arcs are unchanged, the modification to any of the traffic information does not injure the stability of spatial index structure (i.e., R-tree [17]) for road network. Therefore, a kind of queries in ITS application, which refer to the spatial information, can be solved by taking advantages of the spatial index [17]. Another kind of queries, which refer to traffic information, can also be solved effectively. In the next section, we center on solving the second kind of queries by integrating the transportation information with the multi-levels of road map.

3.2.3 Representation of Multi-levels of Transportation Network

Multi-levels of transportation network is an extension of the multi-levels of road network, and is managed by an extended MOR-tree.

3.2.3.1 Integrated Representation of Multi-levels of Transportation Network

Basically, the transportation information is managed on every node of the map. However, the node which connects to road segments in different levels should be processed specially.

Consider the multi-levels of transportation network given in Fig. 3.19. *a1*, *a2* and *b1* to *b3* are basic road segments in *level 1* and *level 2*, respectively. The arcs (or dotted arcs) represent the transportation arcs in *level 1* (or *level 2*). For the simplicity of the figure, the traffic constraints are not drawn in this figure, but the constraints are set as forbidden U-turn, here. MOR-tree for the basic road network in Fig. 3.19 is given in Fig. 3.20. The information about nodes is managed in *level 1* and *level 2* according

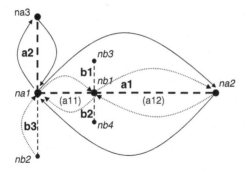

Fig. 3.19 Multi-levels of transportation network

to the level of basic map. The road segments between nodes of different levels can be recorded in the pointer of *nl-ptr* in the composition hierarchy. However, such pointers cannot assure whether there are transportation arcs between these nodes: i.e., *na1* points to *nb1* and *nb2*, but there is no transportation arc from *na1* to *nb2* (there is only an arc from *nb2* to *na1* as *b3* is a one-way road). Consider that the path search in *level 1* is based on the arcs in *level 1* and the search in *level 2* is based on a different arc set. For the search in *level 1*, *Cost-arcs* of node *na1* are:

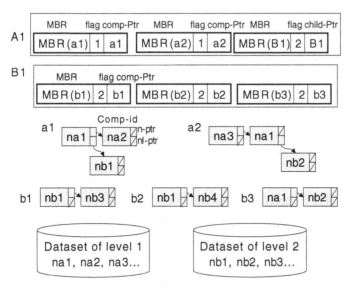

Fig. 3.20 MOR-tree for multi-levels of road networks. *Ai*: internal nodes of main hierarchy; *Bi*: leaf nodes of main hierarchy; *ai*, *bi*: composition hierarchy; *nai*, *nbi*: Composition-entries; *i* (1 or 2): flag of Object-entries or Tree-entries

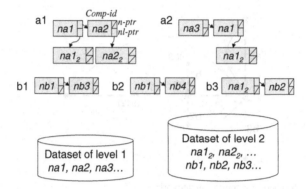

Fig. 3.21 Composition nodes of extended MOR-tree for transportation information

$$\begin{bmatrix} out_{a1}(na_2, cost_1) \\ out_{a2}(na_3, cost_2) \end{bmatrix} \tag{3.16}$$

while, for that in *level 2 Cost-arcs* of node *na1* are:

$$\begin{bmatrix} out_{a11}(nb_1, cost_{11}) \\ out_{a2}(na_3, cost_2) \end{bmatrix} \tag{3.17}$$

The *Constraint-matrices* of node *na1* in *level 1* and *level 2* are:

$$CM(na_1)_1 = \begin{array}{c} \\ in_{a1} \\ in_{a2} \end{array} \begin{array}{cc} out_{a1} & out_{a2} \\ \begin{pmatrix} 1 & 0 \\ 0 & 1 \end{pmatrix} \end{array} \tag{3.18}$$

$$CM(na_1)_2 = \begin{array}{c} \\ in_{a11} \\ in_{a2} \\ in_{b3} \end{array} \begin{array}{cc} out_{a11} & out_{a2} \\ \begin{pmatrix} 1 & 0 \\ 0 & 1 \\ 0 & 0 \end{pmatrix} \end{array} \tag{3.19}$$

Because (1) the information in the lower level (*level 2*) is not to be used by the operation in *level 1* and (2) the original MOR-tree is not consistent with the transportation situation, we split one node in the upper level (*level 1*) into two nodes belonging to two levels: one is the original node in the upper level, another is a transportation-node in the lower level, which contains transportation information in the lower level. In MOR-tree for road network, *nl-ptr* contains a reference to the composition of the parent node object in the lower level. Here, we redefine *nl-ptr* as a pointer which contains a reference to a transportation-node of the parent node object in the lower level. The extended composition hierarchy is given in Fig. 3.21. The nodes *na1, na2* in *level 1* possess the transportation information referred to *level 2*, so the transportation nodes for them are created in *level 2*. For the composition hierarchies of *level 1*, *nl-ptr*'s refer to the corresponding transportation nodes in *level 2*. And, the search operations referred to only the upper level can be performed just by using the original MOR-tree, and the operations referred to the two levels can be performed by using the information managed in the lower level. The nodes which connect only with transportation arcs on the same level can be represented with a *super-node* directly.

3.2.3.2 Spatial Search and Path Search

MOR-tree supports not only the spatial search on multi-levels of maps but also the path search on multi-levels of transportation information. Spatial search to multi-levels of maps can be realized by accessing objects until a specific level via MOR-tree. Since MOR-tree takes the advantages of spatial index structure, spatial queries such as region query can be realized by accessing to the internal nodes and composition hierarchies until a specific level of MOR-tree. The zoom-in/out operations can be realized by accessing the nodes under a specific internal node of MOR-tree.

Path search on the transportation information on different levels can be realized by using an extended "ink-blot" search method. The ordinary "ink-blot" search method solves a path search from v_i to v_t. It begins from expanding v_i's connecting nodes in the sequence of the *cost* on the *cost-arc*. If the target node has not been expanded, the search goes on by expanding the nodes connected by those expanded nodes. For the search based on the multi-levels of transportation network, we extend the method. For example, a path search from city hall A to city hall B in separated prefectures can be realized based on transportation networks on different levels of details. The search can be performed in three steps.

(1) Firstly, the search is based on the city-level of transportation network inside city A using "ink-blot" search method. This step is terminated when a node belonged to the upper level is expanded.
(2) Then, the search goes on in the upper level until a node inside the region city B is expanded.
(3) Finally, the search returns to city level and expands nodes until the target is found.

The correctness of this algorithm is assured by the "ink-blot" method. Here, we leave the proof out. By using MOR-tree, the changeover points between multi-levels are those nodes on the road network which possess original node and transportation-node on multiple levels.

3.2.3.3 Characteristics of *Super-Node* Representation Method

In this section, we compare the features of traffic information represented by our method (denoted as *super-node* method) with those in the method used by [52, 53]

		Without constraints		With constraints	
		Without turn costs	With turn costs	Without turn costs	With turn costs
Node-link method	nodes	m^2	$8m^2-8m$	$4m^2+4$	$8m^2-8m$
	arcs	$4m^2-4m$	$40m^2+12m-4$	$9m^2+8m-5$	$22m^2+4m-2$
Super-node method	Constraint-matrix	Null	Total elements in matrixes: $16m^2+36m-20$		
	nodes	m^2			
	arcs	$4m^2+4m-3$			

Fig. 3.22 Comparison of object numbers managed in different methods under different conditions

(denoted as node-link method). These features lay the foundation for understanding the behaviors of these methods with respect to retrieval and storage.

For a n ($=m \times m$) grid graph which represents a road network with m^2 nodes and $2m(m-1)$ links, Fig. 3.22 gives the number of objects (nodes, arcs and so on) managed in the datasets by using *super-node* method and *node-link* method in different conditions:

(1) Without constraints. The transportation network is specified with travel cost and with/without turn costs;

(2) With constraints. The transportation network is specified with travel cost, traffic constraints, and with/without turn costs. From this table, we can observe that on any condition the number of arcs and nodes managed in the dataset keeps the same by using our method. This is why our method supports the stability of the spatial index of the basic road networks. The stability of spatial index ensures that spatial searches can be realized efficiently and the searches on traffic information can also be performed with a steady cost. Contrary to this, in *node-link* method constraints or turn costs are represented with additional nodes and arcs. When there is no traffic constraint and turn cost on nodes, the traffic arcs can be represented only by the nodes on the road map. When there are traffic constraints, the number of nodes (arcs) is four times (duplicated). When there are turn costs, the number of them is even increased. With the increase, the search cost on traffic information is also increased.

3.2.3.4 Evaluation

We compare *super-node* representation method with the methods used by [52, 53]. We set the experiment environment as follows. The total number of nodes N_{num} is 42,062 and the number of links L_{num} is 60,349 in the basic road map. These road segments are assigned to the country level (country-wide highways and national roads), the prefecture level (prefectural roads and main local roads) and the city level (city roads).

In city level of transportation network, the number of average traffic arcs connecting to a node is about 2.87 ($=2\,L_{num}/N_{num}$). When there is no traffic constraint for the basic road map, in node-link method [53] there are 120,798 records (two times of link numbers in road maps). In this *super-node* method, the amount of information is related to the number of arcs in every node: here, the nodes with four, three, two and one out-arcs are about 24 : 51 : 13 : 12. The total arcs managed in city level by SN method is 120,798. When there are traffic constraints, Right-turn and U-turn are forbidden in about half of the cross and T-junction points. Then, in NL method there are about 142,423 nodes and 135,293 arcs. While in SN method the amount of information keeps the same on any situations.

3.3 Summary

In order to represent map information of GIS smartly, especially for a country-wide integrated GIS, we proposed the Multi-scales/Multi-themes (M^2) map information model, in which map elements are composed as a hierarchical structure with multi-scales so as to deal with various types or different classes of scaled maps successfully. The model is powerful to integrate various scales of maps uniformly in comparison with those in the traditional GIS. Because the map elements are uniquely assigned to a special level on a particular scale, the map elements should propagate the upper levels to lower levels and compute the topological relations between them effectively. The road network is assigned to part of Aichi Prefecture in Japan for the road theme among three different levels in the prototype system with inheritance functions, and compared M^2 model with other two models. Based on those comparisons, the M^2 model outperforms L-model and T-model in management and display of road network.

To achieve an efficient access and maintenance in the road networks with road network model, in the next chapter we concentrate on the index structure for managing road networks.

Chapter 4
Index in Road Network

Moving object data in road network is a complex data type, including not only static spatial road network information, but also the dynamic spatio-temporal information. Therefore, road network query is an operation with a high time and space complexity. Efficient indexing mechanism is an effective solution to improve the efficiency of data access. However, the method in this situation is different from those for objects moving in 2-D free space, because not only the moving directions but also the distance among moving objects are constrained by the road network. Therefore, the neighboring relations among moving objects should take the underlying road network into consideration. R-TPR$^\pm$ tree is a composite index structure used for managing the real-time data of moving objects in the road network. It includes two parts: a static 2-D R^*-tree indexing the road network on the top of a forest of dynamic 1D TPR$^\pm$-trees indexing the moving objects.

To support searching across the city roads, in addition to building an index on moving objects, there also must be different levels of detail or different levels of road network in the index. MOR-tree is created for multi-levels of road networks under M^2 map information model and R-trees are for the datasets in every level managed by L-model, respectively. It can handle spatial data efficiently and provides integrated access to multi-scales of road networks.

To deal with distinct counting of moving object problems in aggregate query, by integrating indexes for network-constrained moving objects (RR-tree) with sketches, Sketch RR-tree manages the sketch synopses of the moving objects in road network considering their "network" positions and supports the effective aggregation on road networks. However, the Sketch RR-tree without provable guarantees on the approximate quality of aggregate queries when the moving objects are in non-uniform distribution. DynSketch uses AMH to partition Sketch RR-tree intelligently and makes vehicles well-distributed in each bucket. Then appropriate number of sketches is assigned for each bucket dynamically so that the quality of aggregate query is improved.

To divide buckets intelligently and make each bucket adapt to the distribution of moving objects automatically, Modified Histogram (MH) is proposed to partition the space and improve the quality of the approximation.

© Springer International Publishing Switzerland 2015
J. Feng and T. Watanabe, *Index and Query Methods in Road Networks*,
Smart Innovation, Systems and Technologies 29,
DOI 10.1007/978-3-319-10789-9_4

This chapter mainly introduces 5 typical index methods which deal with different problems in road network respectively.

4.1 R-TPR$^{\pm}$ Tree

4.1.1 Introduction

With the improvements of geographic positioning technology and the popularity of wireless communication, location-based service (LBS) is proposed to provide the user with a service that is dependent on positional information associated with the user [57]. Prominent examples of LBS concern vehicle navigation, tracking and monitoring, where the positions of air, sea, or land-based equipment such as airplanes, boats and cars are of interest.

This section is concerned with the indexing of moving objects on road network for real time and forecasting purposes: for example, transportation control and real time transportation guidance. It is difficult to index objects moving in 2-D free space, because not only the moving directions but also the distance among moving objects is constrained by the road network. Therefore, the neighboring relations among moving objects should take the underlying road network into consideration.

Past works on indexing of spatio-temporal data concern either past data or present and future data. The former direction includes works in 2-D space [58–60] or in spatial networks [61]. The latter direction includes works on indexing the present positions of moving points [2, 59, 62–65], all these approaches deal with spatial data in 2-D space but cannot be applied to the moving objects in spatial network.

A straightforward method for indexing the static road network and the moving objects at the same time is the composite structure, shown in Fig. 4.1. The composite structure is made up of two parts: one is a static tree managing the information of road network, whose leaf node like R1 in Fig. 4.1 denotes a region of the road network, pointing to a dynamic tree indexing moving objects move in this region, and all these

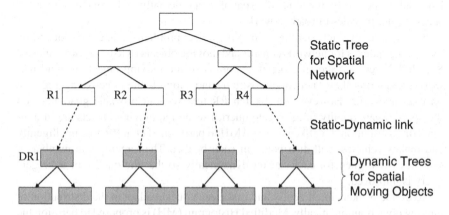

Fig. 4.1 Image of a composite tree

dynamic trees compose the dynamic index under the static tree. The other is using composite structure, we can insert and query spatial locations efficiently with the static tree, and update the positions of moving objects quickly as we divide them into small subsets and index them separately.

However, to ensure the effect of the composite structure for indexing current positions of moving objects on road network, there are two issues to be solved:

- **Update cost** The update cost includes the cost for one operation and the times of operations. In current composite structures, index element of the static tree is road segment (in other words, the forecasting region is limited by the segment), and results in frequent updates when the objects move into other segments. Therefore, the issue of enlarging the forecasting region and decreasing the update times is the key to a "really" efficient composite tree.
- **Query cost** The queries on moving objects include the region query, nearest neighbor query, aggregate query and so on. In applications, aggregate queries are important to answer such kinds of questions: whether a specific road segment is fluent, or what is the average speed on a road? Therefore, how to make up an index structure to support kinds of queries should be considered.

In this section, methods are proposed to get an efficient composite structure for indexing moving objects on road network. We bring in the idea of road connection algorithms [66] for dividing the approximate regions of static tree, which uses the road network within which the objects are assumed to move for predicting their future positions, and can increase the forecasting period of moving object and decrease the update times of the database. Furthermore, we propose new spatio-temporal index structure for moving objects in every road region, which arranges the objects by their moving directions as well as their locations. This structure supports the aggregate and spatio-temporal queries more efficiently than other structures.

4.1.2 Road Connection Algorithms

There is reality meaning in connecting several road segments as an index element. On one hand, regarding the connected segments as a group will decrease the element number of the static tree, and increase the query efficiency. On the other hand, the relativity of road segment provides the important cause for segment selection for a moving object when it faces to several segments at a cross node. The better the connection algorithm is, the more precise the forecasting is. However, how to decide the size of forecasting region or the length of forecasting path is a problem in monitoring moving objects on road network. When the length of forecasting path is too long, the forecasting precision cannot be approved; while it is too short, the gathering of sample locations will be a heavy burden.

To find out a better way for road connection, LSC algorithm (length and side-based segment connection) and ASC algorithm (angle-based segment connection) [66] are applied to the real road networks (California Road Network from Geographic Data Technology Inc. 1999). Experiments show that the number of road segments is decreased more than 59 %, and that the length of segments and the number of nodes are increased. The road networks processed by ASC is better for representing the real road network for car tracking, because the number of main roads is less than those non-main roads, and the nodes included in the main roads are more, and length is longer than those of the non-main roads. These properties are just similar to the driving properties of cars in road networks.

In the composite tree for the moving objects in road networks, the problem of creating a "good" static tree for road network is how to decrease the update times occurred at every changing segment time of moving objects, and also the segment connection algorithms just can bring a practical way for solving this problem. In the followings, the road segment processed by ASC algorithm is used as the basic index element in the static tree.

4.1.3 Framework and Query Method

R-TPR$^{\pm}$ tree is a composite index structure used to manage the real-time data of moving objects on the road network. It is composed of two parts: a static 2-D R^*-tree indexing the road network on the top of a forest of dynamic 1-D TPR$^{\pm}$-trees indexing the moving objects. Objects on every road segment are supposed to be moving at just the same speed as that at the sampling time. Though the directions of them may change all the time along the road segments, if they are still in one segment, it is thought to be keeping the same direction. Therefore, in a road segment, the moving directions of moving objects are just two: from start node s to end node e (denoted as + direction), or from e to s (denoted as—direction). Furthermore, in the real road networks, the cost or weight of a segment in two directions is not always the same; a TPR$^{\pm}$-tree is proposed for indexing the spatio-temporal information in one road segment.

TPR$^{\pm}$-tree is a tree with a root node and two subtrees. In every subtree, only objects in the same direction are managed; in other words, the two subtrees are managing the objects in different moving directions. In the root node, there are two pointers for these two subtrees: one is +, and the other is −. Figure 4.2 depicts this situation. In more formal speaking, there are two elements of R-TPR$^{\pm}$ tree's data set, R and O.R is the road network and O is the moving objects. After dividing R into segments by ASC algorithm, there will be a set of independent segments, R = $\{S_1, S_2, ..., S_n\}$, where $S_i(1 \leq i \leq n)$ means a segment of the road network, and $\leq n$ is the number of the segments in R. While R is divided, the set of moving objects O is also divided into n subsets, represented as O = $\{O_1, O_2, ..., O_n\}$, and O_i = $\{mo_p, mo_{p+1}, ..., mo_q\}(1 \leq i \leq n$), where mo_j ($p \leq j \leq q$) denotes a moving object that moves in the segment S_i, and we will get: NUM(O) = $\sum_{i=1}^{n} NUM(O_i)$, where NUM(O_i) means the amount of moving objects in the subset O_i. As we said before, in each segment of the road network, a moving object moves from the start point to the end, or opposite. We use TPR$^+$-tree (TPR$^-$-tree) to index the data of the set O_i in +(−) direction. The leaf node of the TPR$^+$-tree (TPR$^-$-tree) is denoted as

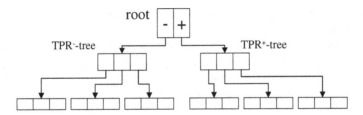

Fig. 4.2 Image of TPR$^{\pm}$ tree

(j-Identifier, P_j, V_j), where j-Identifier refers to the object O_j. $P_j = \frac{Length(mo_j, A)}{Length(A, B)}$ is the relative position of the segment, where $\leq A$ is the start point of the segment, and $\leq B$ is the end point, Length($\leq A$, $\leq B$) means the length between $\leq A$ and $\leq B$. P_j approaches to 0 means that O_j is as close to the start point $\leq A$ as much, and if O_j is more close to the end point $\leq B$, P_j will approaches to 1 as well. At the same time, V_j means the velocity of O_j, when O_j moves from $\leq A$ to $\leq B$, then $V_j > 0$, and $V_j < 0$ when it moves opposite. The non-leaf node of this TPR^{+}-tree (TPR^{-}-tree) is denoted as (P_s, P_e, V_s, V_e, child-point). Where child-point is the address of a lower node, $|P_s, P_e|$ covers the position interval of its lower node's entries, and $|V_s, V_e|$ covers the velocity interval. As there are $\leq n$ subsets of O, there will be $\leq n$ TPR$^{\pm}$-trees, and we can use a forest for indexing the moving objects in the road network.

Algorithm 1: Forecasting traffic Flow FTF(R, t)

    input : R is the predict range, and t is the forecasting period.
    output: $Flow[]$ contains the traffic flows every segment will have after t seconds.
1 **begin**
2   int $Flow[]$;
3   Initialize a node array $TransArray$;
4   TPR$^{\pm}$-tree $trees[]$;   /* find those TPR-trees inside R */
5     $trees$ = 2-D R*-tree.RangeQuery(R);
6   **while** $trees$ is not empty **do**
7        $tree$ = trees.headNode();
8        $trees$.removeHead;
9        $Flow[tree$.ID] = Single Segment Forcast($tree.^+node, t$);
10       $Flow[tree$.ID] = Single Segment Forcast($tree.^-node, t$);
11   **while** $TransArray$ is not empty **do**
12         $tree$ = $transArray$.headNode.selectRoute;
13         $tree$.Insert($transArray$.headNode);
14         $Flow[tree$.ID] = Single Segment Forcast($tree.^+node, t$);
15         $Flow[tree$.ID] = Single Segment Forcast($tree.^-node, t$);
16         $tree$.delete($transArray$.headNode);
17         $transArray$.removeHead;
18   **end**
19   **end**
20 **end**

Segments R are indexed by a 2-D R^*-tree. The leaf node of the index is denoted as: (i-Identifier, MBR_i, TPR$^\pm$-point, P_i). Where i-Identifier refers to the segment S_i, MBR_i is a minimal bounding rectangle of the segment, TPR$^\pm$-point is the pointer to the TPR$^\pm$-tree indexing the objects of O_i, and P_i is something about other properties of the segment, such length. The non-leaf node of this R^*-tree is denoted as (child-MBR, child-pointer), where child-pointer points to a lower node, and child-MBR covers all MBRs of its lower nodes.

Algorithm 2: Single segment forecast SSF(N,t)

---

    input : N is a node of TPR tree, at the beginning of this algorithm, it means the root node of
          TPR^+-tree (TPR^--tree); t is the forecasting period.
    output: $Flow$ is the number of cars which will be still in the segment t seconds later indexed
          by N

```
 1 begin
 2 │   int Flow;
 3 │   if NPe ≤ 1 then
 4 │   │       /* all cars indexed by N will not exceed the segment t
       │           seconds later */
 5 │   │   Flow += N.carNum;
 6 │   else if NPs > 1 then
 7 │   │       /* all cars indexed by N will exceed the segment t
       │           seconds later */
 8 │   │   foreach subNode Ni in N do
 9 │   │   │   TransArray.add(Ni);
10 │   │   end
11 │   else if N.level == 0 then
12 │   │       /* part of cars will be in the segment and N is a leaf
       │           node   */
13 │   │   foreach subNode Ni in N do
14 │   │   │   if Moi.postion+Moi.veloctiy ≤ 1 then
15 │   │   │   │   Flow ++;
16 │   │   │   else
17 │   │   │   │   TransArray.Add(Moi);
18 │   │   │   end
19 │   │   end
20 │   else
21 │   │   foreach subNode Ni in N do
22 │   │   │   Flow += flowCal(Ni);
23 │   │   end
24 │   end
25 │   return Flow;
26 end
```

In application, it is more important to get aggregate information than to calculate exact positions of cars moving in the road network. A forecasting aggregate query Q is denoted as (R, $\leq t$), where R describes the forecasting range, and $\leq t$ refers to the forecasting period. Using the 2-D R^*-tree, we can find out the segments which we want to calculate inside R, and those TPR$^\pm$ trees indexing moving objects in these segments. And we get how many cars will be still in this segment after $\leq t$ seconds later according to their current speeds and positions with this forecasting method.

The non-leaf node of the TPR$^\pm$ tree is denoted as (P_s, P_e, V_s, V_e, child-point). As the moving directions of those objects indexed by the same node are the same, we assume their speeds keep constant. We can get:

$$NP_s = P_s + V_s * t \qquad (4.1)$$

$$NP_e = P_e + V_e * t \qquad (4.2)$$

Here, $|NP_s, NP_e|$ covers the position interval which those objects will not overstep after $\leq t$ seconds. So if NP_e is less than 1, then all those objects indexed by this node will not exceed this segment $\leq t$ seconds later, and if NP_s is larger than 1, then all those objects will leave this segment $\leq t$ seconds later. When NP_s is less than 1 and NP_e is larger than 1, recurs is needed to its sub-nodes and repeat this algorithm until arriving to the leaf nodes.

Those objects leave the segment after $\leq t$ seconds will select proper segments, insert into those other TPR$^\pm$ trees, and delete after the algorithm ends. Those objects are added into the traffic flow of current segment.

4.1.4 Evaluation

For the experiments, a PC of Pentium IV CPU with 3.19 GHz and 512 MB of main memory was used. The trees and the algorithms were implemented in C++ and compiled using Visual C++ version 2003.

The moving objects in a road network are indexed by R-TPR$^\pm$ Tree and R-TPR-tree (the Static Tree is R*-tree and the Dynamic Tree is TPR-tree). The validity tests and tree node disk access tests are carried out for the vehicles moving on a 1 Km road segment. The results of the queries varied with different vehicle numbers, vehicle speed and forecasting period. The number of disk accesses of our method was quite fewer than that of TPR-tree, and it was a stable. And it was clear that the validity of this method is better than that of TPR-tree while the forecasting period is shorter.

4.2 MOR-Tree

4.2.1 Introduction

Under M^2 map information model, map objects of a theme are assigned to multi-levels of datasets without repetition. For a particular requirement, map information can be prepared dynamically by using inheritance and overlay functions.

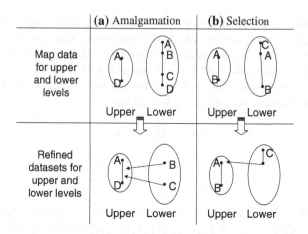

Fig. 4.3 Refinement

Typical situations of this assignment for road objects in two levels are given in Fig. 4.3. In Fig. 4.3a, the relation between the map objects in upper and lower levels is amalgamation: the map object AD is represented as a line (AD) in the upper-level map and as lines (AB, BC, CD) in the lower level—an object in the upper level can be regarded as the merging of adjacent objects in the lower level. The information about nodes A and D is the same in both levels but the relation between A and D is different. So, under our model the information in the lower-level map can be split into two parts: one part is the same information as that in the upper level, and the other part is the information which belongs to only the lower level. These two parts are called *refined datasets*, and are managed in the upper and lower levels, respectively. Considering the relations between two datasets are only needed in generating maps of the lower level, these relations are managed in the refined dataset of the lower level. In Fig. 4.3b, there is a selection relation between maps in two levels: the map object AB is displayed in two levels, but AC is not needed in the upper level—the information in the upper level is a "selection" of that in the lower level. Therefore, the information about the refined datasets in the upper and lower levels is the objects AB and AC, respectively.

4.2.1.1 Methods for Accessing Refined Datasets

There are at least two kinds of users in manipulating map datasets. One is an end-user, who wants to get the latest maps for desired region, themes and level, and makes spatial queries on maps. The other is an administrator, who is a manager of maps and takes the responsibility of map maintenance. So, we propose two methods for accessing refined datasets (Fig. 4.4): one is to make use of functions to generate maps for end-users, and the other is to use an index structure to access the refined datasets directly.

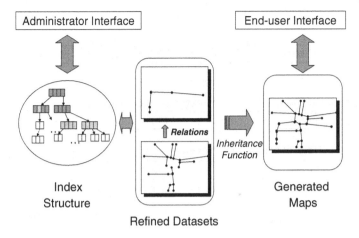

Fig. 4.4 Two methods for accessing refined datasets

4.2.2 Index Structure

MOR-tree is proposed for achieving the following abilities: (1) the ability of handling spatial data efficiently; (2) the ability to provide integrated access to multi-scales of maps; and (3) the ability of arranging the relations among multi-levels of maps.

Besides the ability of accessing spatial objects in one level (in one dataset) efficiently just like other spatial index structures, MOR-tree is designed to be able to differentiate the levels of objects and manage the relations among objects of multiple levels by adopting two kinds of hierarchies.

A main hierarchy is proposed to differentiate the levels of road objects by assigning logical importance value to every object. The logical importance value is a natural number in agreement with the map level: e.g., 1 for objects in country-level datasets, 2 for those in prefecture-level datasets, and so on. The objects with the higher importance (with the smaller importance value) are stored in the higher levels of the hierarchy. Another kind of hierarchy, called a composition hierarchy, is proposed to keep the relations among levels, which are pointed by the leaf nodes of the main hierarchy. The main hierarchy is based on R-tree, to achieve the outstanding spatial access performance. Each node in MOR-tree contains a number of entries. There are three kinds of entries: tree-entries, object-entries and composition-entries. The internal nodes may contain the first two kinds of entries, in contrast to R-tree. The leaf nodes contain object-entries. The object-entry points to a composition hierarchy, which consists of composition-entries (Fig. 4.5).

These kinds of entries have the following forms:

(1) Object-entry has the form (*MBR*, *flag*, *comp-ptr*), where *MBR* is the minimal bounding rectangle in the composition hierarchy; *flag* is a natural number that indicates the importance level; and *comp-ptr* contains a reference to a composition hierarchy.

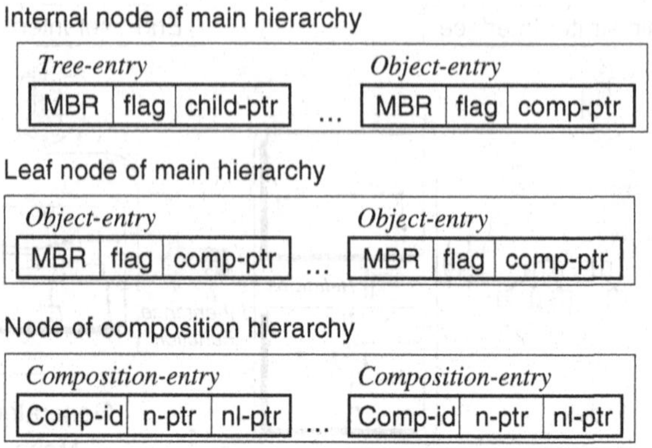

Internal node of main hierarchy

Tree-entry				Object-entry		
MBR	flag	child-ptr	...	MBR	flag	comp-ptr

Leaf node of main hierarchy

Object-entry				Object-entry		
MBR	flag	comp-ptr	...	MBR	flag	comp-ptr

Node of composition hierarchy

Composition-entry				Composition-entry		
Comp-id	n-ptr	nl-ptr	...	Comp-id	n-ptr	nl-ptr

Fig. 4.5 Nodes in MOR-tree

(2) Tree-entry has the form (*MBR, flag, child-ptr*), where *child-ptr* contains a reference to a subtree. In this case *MBR* is the minimal bounding rectangle for the whole subtree and *flag* is the smallest importance value for the child-nodes.

(3) Composition-entry has the form (*comp-id, n-ptr, nl-ptr*), where *comp-id* is the identifier of object's composition; *n-ptr* contains a reference to the next composition of the parent node object: e.g., when the object is a road segment, the first composition of it is one of the end points, and *n-ptr* points to the next point on the same object; and *nl-ptr* contains a reference to the composition of the parent node object in the lower level: e.g., the intersection between the upper-level road and lower-level road.

Here, object-entry and composition-entry are given in Fig. 4.6 as an example. In Fig. 4.6a, there is a link object *a1* with compositions *na1* and *na2* in *level 1*, so the entry in the index structure for *a1* is an object-entry, which consists of *MBR* for *a1* (for simplicity, *a1* is used to replace *MBR* in this figure), *flag* of *level 1* (here it is *1*) and *comp-ptr* which points to one composition of *a1* (here it is *na1*). The compositions of *a1* (*na1, na2*) are managed by composition-entries. *n-ptr* of *na1* points to the entry for *na2*, and the other pointers of two composition-entries are NULL. A more complex example is given in Fig. 4.6b, in which there are lower level compositions (*nb1* and *nb3*) of *a1*. The objects *b1* and *b2* belong to *level 2*. So, *nl-ptr* of *na1* points to the composition-entry *nb1* of *level 2* and *n-ptr* of *nb1* points to *nb3*.

4.2.3 Algorithms for Operations of MOR-Tree

Because the MOR-tree is created not only based on the spatial information of map objects but also the levels of them. A new object-entry should be inserted to the main hierarchy on a specific level of the MOR-tree. The main step for the insertion is given in Algorithm 3.

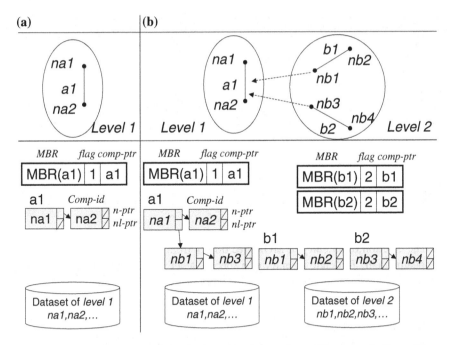

Fig. 4.6 Example of MOR-tree for two-levels of road networks. **a** Object-entry. **b** Composition-entry

---

Algorithm 3: Insertion(E)

input : an object-entry E with *Flag* of level and its composition hierarchy
output: updated MOR-tree – *MOR-T*
1 **begin**
2    $L =$ ChooseLeaf(*MOR-T, Flag*);   /* choose leaf on a specific level */
3   **if** $L.hasRoom()$ **then**
4 | $L.add(E)$;
5   **else**
6 | *SplitNode(Flag)*;   /* to obtain L and LL */
7   **end**
8   *AdjustTree(L, LL)*;
9   **if** *split propagation caused the root to split* **then**
10 | *Create(MOR-T)*;
11   **end**
12 **end**

Choosing the leaf node on a specific level of main hierarchy is similar to that of choosing the leaf node in the R-tree. However, the leaf of a specific level may be the internal node of main hierarchy in the MOR-tree. Therefore, a node in main hierarchy may be a leaf for a specific level or be an internal node for another level. The main step in ChooseLeaf is given as Algorithm 4.

Algorithm 4: ChooseLeaf(N, $Flag$)

---

input : an object-entry E with $Flag$ of level and its composition hierarchy;
output: the node N on the level $Flag$

1 **begin**
2   **if** *isLeaf (N, Flag)* **then**
3 | **return** N;
4   **else**
5   Choose the node Nc from $N's$ children whose increasing area of MBR is smallest
        when inserting E into it;
6       $N = Nc$;
7   ChooseLeaf (N, $Flag$);
8   **end**
9 **end**

The deletion of an entry in MOR-tree differs from that in R-tree on the process of locating leaf. To delete an entry of a specific level should first find the leaf node for the specific level, and then remove the entry and condense main hierachy.

Algorithm 5: Delete(E,$Flag$)

---

input : an object-entry E with $Flag$ of level and its composition hierarchy;
output: updated MOR-tree – *MOR-T*

1 **begin**
2   L = FindLeaf(*MOR-T*, $Flag$); /* find leaf including E on a specific level */
3   L.Remove (E, $Flag$);
4   CondenseTree( );
5   **if** *the root node has only one child* **then**
6 | Make the child the new root;
7   **end**
8 **end**

The algorithm for finding the leaf node (the specific entry is in) is given in Algorithm 6.

4.2.4 Indexing Process for Two-Level Road Networks

The index is created based on datasets of multiple levels. When a map is generated for a region on a specific scale, the indexes are created on the referred datasets. Here, the main steps are given in the creation of MOR-tree datasets from two levels:

(1) Create bottom nodes of main hierarchy based on the objects in the lower level.

Algorithm 6: FindLeaf(T, $Flag$)

input : T:The MOR-T Tree,$Flag$: The flag of level
output: an object-entry E with $Flag$ of level and its composition hierarchy
1 **begin**
2   **if** $T \neq Leaf(Flag)$ **then**
3   **if** *MBR of E overlaps with MBR of T's child entry Tc* **then**
4   FindLeaf (Tc, $Flag$);
5   **end**
6   **else if** $T == E$ **then**
7   **return** T;
8   **end**
9 **end**

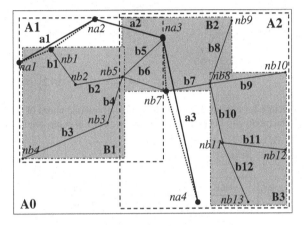

Fig. 4.7 Example of two-level of road networks and MBRs for road segments

(2) Create composition hierarchy for objects from the upper level: *nl-ptr* points to the intersection node, which is the intersection between upper-level and lower-level roads.
(3) Create main hierarchy based on the nodes generated in (1) and (2).

An example for two-level of road networks is given in Fig. 4.7. *a1* is the road segment managed in the upper level. The bold lines (*a1*, *a2* and *a3*) represent the road segments managed in the upper level, and lines *b1–b12* represent the road segments managed in the lower level. The dotted lines between *na1* and *nb1*, *nb1* and *na2* are generated by the inheritance function, which propagates *a1* to the lower level and splits the original road segment into two segments. The dotted lines between *na3* and *nb7*, *nb7* and *na4* represent the road segments generated by inheritance function for *a3*. The index for road datasets in Fig. 4.7 is given in Fig. 4.8, the detail contents inside every node are given in Fig. 4.9, and *MBR*'s generated for the index are illustrated with dotted boxes in Fig. 4.7 (for simplicity, in the example the maximum number of entries in a main-hierarchy node is 4). At the step 1, bottom nodes *B1–B3* are generated based on the objects in level 2 (the gray boxes in Fig. 4.7); At the step 2,

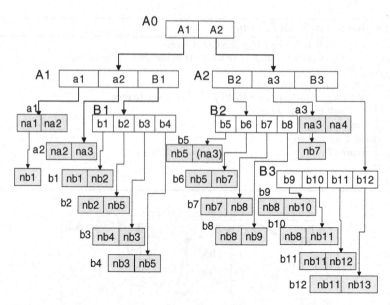

Fig. 4.8 Index structure for two-levels of road networks. *Ai* internal nodes of main hierarchy; *Bi* leaf nodes of main hierarchy; *ai, bi* composition hierarchy; *nai, nbi* composition-entries

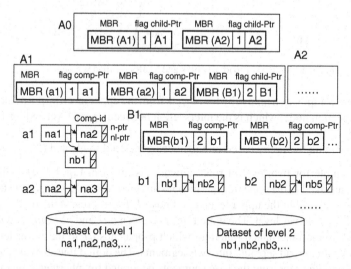

Fig. 4.9 Detail contents inside nodes of MOR-tree for two-levels of road networks. *Ai*: internal nodes of main hierarchy; *Bi* leaf nodes of main hierarchy; *ai, bi* composition hierarchy; *nai, nbi* composition-entries; *i* (1 or 2) flag of object-entries or tree-entries

the composition hierarchies corresponding to objects *a1*, *a2*, and *a3* in level 1 are created. For example, the composition hierarchy for *a1* contains three *composition-entries* of *na1*, *na2* and *nb1*. *comp-id* of *na1* (*na2*) points to the real disk page which node *na1* (*na2*) is located in the dataset of level 1, *n-ptr* and *nl-ptr* of *na1* point to entries for *na2* and *nb1*, respectively. *comp-id* of entry *nb1* points to the real disk page which node *nb1* is located in the dataset of level 2. *MBR* for every object-entry is based on all the objects inside the corresponding composition hierarchy. At the step 3, the main hierarchy is created based on the bottom nodes (*B1*, *B2*, and *B3*) and the objects in level 1 (*a1*, *a2*, and *a3*). The creation method is similar to that of R-tree. Here, first the inner-nodes *A1* and *A2* are created, which point to (*a1*, *a2*, *B1*) and (*a3*, *B2*, *B3*), respectively. And then *A0* is created as the root of MOR-tree.

4.2.4.1 Example of Update for Two-Level Road Networks

There are flags in object-entries and tree-entries used to mark out different levels. Moreover, the relations among levels are represented in the composition hierarchies and the simultaneous modification at multiple levels is possible.

Consider an example which is related to the deletion of the road segment between *nb1* and *na2*. The operation refers to datasets in two levels. Before the update, *nb1* belongs to the level 2, and *na1* and *na2* belong to the level 1. The road segments *na1–nb1* and *nb1–na2* are generated in the level 2; however, after the update, as the node *nb1* becomes an end-point of the upper-level road, the road segment managed originally in the upper-level $\leq$ *a*1 (*na1–na2*) should be replaced by *na1–nb1*, and the information about *nb1* should be managed in the upper level. The update steps are given as follows:

(1) Descend the main hierarchy, and locate the referred object-entry of object in the upper level. Also, locate modification to composition hierarchy, and replace *na2* with *nb1* in composition hierarchy (see Fig. 4.10).
(2) Modify corresponding datasets. Next, copy node *nb1* from level 2 to level 1, and make up the original one as a virtual node, which points to the real one in level 1.
(3) Adjust the main hierarchy.

4.2.5 Evaluation

4.2.5.1 Characteristics of MOR-Tree

In this subsection, the features of MOR-tree are compared with that of R-tree. Here, MOR-tree is created for multi-levels of road networks under M^2 map information model and R-trees are for the datasets in every level managed by L-model, respectively. These features lay the foundation for understanding the performance of two methods: the implementation cost of indexes; the superiority in MOR-tree holds for different operations such as the integrated maintenance of multi-levels of map datasets.

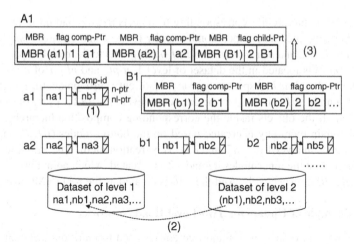

Fig. 4.10 Update on index structure. (1) replace *na2* with *nb1*; (2) copy information of *nb1* from level 2 to level 1; (3) reset pointers and adjust main hierarchy

To simplify the analyses, grid graphs are used. Suppose that there are a N_{r1} grid for representing the road segment objects managed in level 1 and a N_{r2} grid in level 2. There are k pairs of objects in different levels with amalgamation relations (defined in Sect. 3.1). That is to say, there are N_{r1} real road segments in the road map on level 1. However, by using inheritance function defined in M^2 model, N_{r1} road segments are extended to $N_{r1} + k$ road segments in level 2, and the total number of real road segments in road map on level 2 is $N_{r1} + k + N_{r2}$.

MOR-tree arranges the objects managed in level 2 and those in level 1 for different levels of main hierarchy, and arranges the relations between objects for composition hierarchies. As the possible height of composition hierarchies is equal to the number of objects' levels, the height of MOR-tree is at most $\lceil \log_m (N_{r1} + \lceil N_{r2}/m \rceil) \rceil$ (m is the minimum number of entries in one node of tree structure). On the other hand, if R-tree is used to arrange the road segments in two levels, two structures should be used respectively, and moreover, as there is no structure for processing the relations between two levels, R-tree should arrange all the real road segments in every level (the objects managed in every level by L-model) and the number of real road segments in level 2 is $N_{r1} + k + N_{r2}$. Therefore, the height of R-tree in level 2 is $\lceil \log_m (N_{r1} + k + N_{r2}) \rceil - 1$ and that in level 1 is $\lceil \log_m N_{r1} \rceil - 1$.

When there is a maintenance requirement referred to objects in two levels, objects from different levels arranged by MOR-tree can be processed directly by traversing the same MOR-tree. However, those arranged by R-tree should be processed respectively by traversing two R-trees.

4.2.5.2   Evaluation of Index Structure for Maintenance

To evaluate MOR-tree, we compare the performance of this index structure with that of multiple datasets (datasets for multiple levels of road networks in L-model) arranged by R-tree in every level, respectively. The comparisons are based on the time of creating index structures for multi-levels of datasets and the modification time for multi-levels of datasets based on these index structures. As the performances are related to the quantity of information in dataset and its distributions among multiple levels, the experiments are based on varied datasets.

The tree structures of MOR-tree and R-trees in test are kept in the main memory, and the information about nodes on road network is stored into the disk pages of the secondary memory. The workstation has 64 MB of main memory and 80 GB of disk. The page size is 4 KB for disk I/O. In MOR-tree, the size of each record $(comp - id, coordinate, n - ptr, nl - ptr)$ for node object in *composition-entry* is 40 Bytes $(8 + 8 * 2 + 8 + 8)$, which leads to 100 records per disk page. In R-tree, the record for node object is $(nd - id, coordinate)$, whose size is 24 Bytes $(8 + 8 * 2)$ and leads to 170 records per disk page.

(1) *Creation time for index structures*: To create an MOR-tree index structure for the road datasets in our prototype system, the road objects in the lower level are arranged in the bottom level, and the objects in the upper level are arranged in the upper level of the tree. The Sort-Tile-Recursive algorithm is used for packing the trees. This algorithm clusters rectangles in an attempt to minimize the number of nodes visited while processing a query. The main part of the creation time comes from the time for sorting MBR's of every tree level. The cost of this sort is $\Theta(n^2)$, so the sort cost of MOR-tree is $\Theta(n_{r1}^2 + n_{r2}^2)$ and sort cost of the two R-trees are $\Theta(n_{r1}^2)$ and $\Theta(n_{r2}^2 + k^2)$. The difference between the two methods becomes larger with the increase of k. k is the number of increased objects in the lower level for those objects managed in the upper level. This means that when more objects in the upper level respond to several objects in the lower level, MOR-tree creation time is faster than that of multiple R-trees. However, the larger k is, the more the composition hierarchies in MOR-tree should be created. Considering the creation time of composition hierarchies for MOR-tree, the comparison of creation time of two methods becomes complex. It shows that the creation time of MOR-tree is shorter than those of R-trees for all test datasets, and the increase of k results in a faster sorting process for creating main hierarchy of MOR-tree than those of R-trees, but also leads to more cost for creating composition hierarchies of MOR-tree.

(2) *Modification to multi-datasets*: As the relations among multiple levels are managed by MOR-tree, the modification to multi-levels of datasets can be realized integratedly without searching the relations among them from the datasets. A "delete" operation referred to two levels of datasets is executed. It shows that the execution time via MOR-tree is shorter than that via R-tree. As the structures are all stored in the memory, the difference of execution time is nearly static. However, when the structures become too large to be stored in the main memory, the R-tree method needs more disk-accesses for finding out the relations among levels, while by using MOR-tree, the relations are managed in composition hierarchies and the times of disk-access can be decreased.

4.3 Sketch RR-Tree

In Intelligent Transportation Systems (ITS), the moving positions gathered from the road network-constrained vehicles are a kind of data stream. How to process queries on the traffic data streams is important for transportation control: e.g., the retrieval of summarized information about moving objects that lie in a query region during a query interval ("how many vehicles have passed through Tian'anmen Square from 8:00 to 9:00 this morning?").

The straightforward method to answer this kind of aggregate query is to access every single record qualifying the query, as suggested in [67]. However, there are motivations for specialized aggregation methods according to [68]. For example, in ITS, (1) the individual data (positions of vehicles in road network all the time) is highly volatile and involves extreme space requirements, while the aggregate information is usually more stable over long periods, thus requiring considerably less space for storage; (2) the query usually requires distinct spatio-temporal count answer. A distinct spatio-temporal count (DC, for short) query q (with window q_r and time interval q_t) returns the number of matching objects. In the situations of data stream, if n is the number of distinct objects and T is the total number of time stamps in history, any solution that solves distinct counting queries precisely needs $O(n \times T)$ space. This is prohibitive in practice since both n and T may be very large.

There are approximate methods for cutting down the data storage by extracting the approximate values from the individual data, such as adaptive mulit-dimensional histogram [69] for approximate query of current data, and method of using several multi-tree structures based on R-tree [17] and B-tree (aRB-tree [67]) for history data. The shortcoming of that approach is the so-called distinct counting problem (if an object remains in the query region for several time stamps during the query interval, it will be counted multiple times in the result). To solve the distinct counting problem, Tao et al. [68] proposed method by integrating R-tree and B-tree with sketches, traditionally used for approximate query processing. Their method is a modification of approximate index aRB-tree [67] which cannot solve distinct counting problem. However, their method can only be appropriate for 2-D free moving objects. In ITS, we should pay more attention to the aggregate information of road network-constrained moving objects on road segments.

We extends Tao's method by integrating indexes for network-constrained moving objects (RR-tree [70]) with sketches. Experiment shows that aggregate queries to get approximate results can be attained when using a proper number of sketches.

4.3.1 Sketch and Sketch Index

The directly-related previous work is sketch index proposed by Tao et al. [68], which solves the distinct counting problem in district aggregate query. The start point of their method is to use the FM algorithm of sketch discussed in [68] to solve this problem. Their method is a modification of approximate index aRB-tree [67] which cannot solve distinct counting problem.

4.3.1.1 Sketch

Typical sketch algorithm for distinct approximate counting is FM_PCSA algorithm [71]. FM_PCSA algorithm was developed by Flajolet and Martin using *Probabilistic Counting with Stochastic Averaging* (PCSA) (referred to as FM_PCSA in the sequel). A sketch (traditionally used for approximate query processing) consists of r bits, whose initial values are set to 0. FM_PCSA uses m independent sketches, each with its own independent hash function h, as in Eq. (4.3). It takes the id of an object o (e.g., the id of a vehicle) as inputs, and a pseudo random integer $h(o)$ with a geometric distribution as outputs.

$$Prob[h(o) = v] = 2^{-v}, \text{ for } v \geq 1 \qquad (4.3)$$

For every object o in the dataset, FM_PCSA applies another hash function to choose one of the m sketches. Then, FM_PCSA sets the $h(o)$-th bit of that sketch to 1. After processing all objects, FM_PCSA let $k_1, k_2, \ldots, k_m$ be the positions of the first 0 in the m sketches respectively. As a result, the number of distinct objects n is estimated as:

$$n = 1.29\, m \times 2^{k_a}, \text{ where } k_a = \frac{1}{m} \sum_{i=1}^{m} k_i \qquad (4.4)$$

4.3.1.2 Sketch Index Structure

The sketch index structure [68] integrates R-tree [17] and B-tree with sketches generated by FM_PCSA algorithm. Figure 4.12 illustrates the sketch index structure for the example of Fig. 4.11. In Fig. 4.11a, assume the set of regions $r_1, r_2, \ldots, r_4$ and consider that at each timestamp, the sketches in the regions are given in Fig. 4.11b. For each region r_i $(1 \leq i \leq m)$ and timestamp t, they maintain a sketch $s_i(t)$ that captures the ids of objects in r_i at t.

In Fig. 4.12, an R-tree indexes the regions $r_1, r_2, \ldots, r_4$. Each R-tree entry is associated with a B-tree. The B-tree records the historical sketches of the corresponding region (or regions in its sub-tree). For example, the B-tree of region r_1 consists of 4

(a)   **(b)**

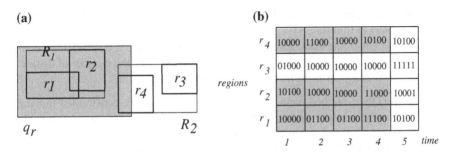

Fig. 4.11 Regions and their conceptual sketch storage model. a Regions. b Scketch

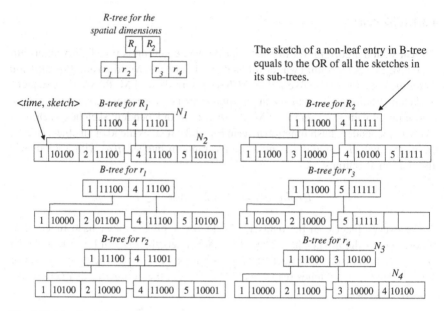

Fig. 4.12   The sketch index structure

leaf entries (in the format <time, sketch>), indicating its 4 sketch changes in history (i.e., no change at time 3). The sketch 11100 of the first root entry in this B-tree equals the OR of all the sketches (i.e., 10000, 01100) in its subtree. The same rule applies to all intermediate B-tree entries and R-tree entries. Consider the first leaf entry <1,10100> in the B-tree of R1. Its sketch 10100 equals the OR for sketches of r_1 and r_2 (i.e., 10000, 10100, respectively) at time 1. For each intermediate R-tree entry R_i, whose sketch at any time t, is the OR of sketches of all the regions in the subtree (of R_i) at t. In general, for each intermediate R-tree entry R_i, whose sketch at any time t, is the OR of sketches of all the regions in the subtree (of R_i) at t. The sketch index is a dynamic structure, and its incremental maintenance algorithms follow those of the aRB-tree due to the similarity of the structures.

Based on the following two Lemmas, they proposed methods for district distinct counting and distinct sum queries.

- Using the sketch $s_i(t)$ in cell (r_i, t), we can estimate the distinct number $n_i(t)$ of objects in r_i at timestamp t.
- Using the OR of the sketches of c cells $(r_{x1}, t_1), (r_{x2}, t_2), \ldots, (r_{xc}, t_c)$, (two cells may have the same region or timestamp), we can estimate the number of distinct objects o that appear in some region r_{xj} at time t_j.

However, their method cannot be applied to road network-constrained moving objects directly, because the sketches are computed for the regions of R-tree. In the situation for road segment aggregate query, the aggregate values is usually for segment (segments). In their sketch index structure, there are no recordings of those data.

4.3.2 RR-Tree for Road Networks

A road network with nodes and polylines representing respectively the crosses and road segments can be regarded as an undirected graph G, $G = (V, PL)$, where V is a set of vertices $\{v_1, v_2, ..., v_n\}$, and PL is a collection of polylines $\{pl_1, pl_2, ..., pl_m\}$. pl_i is a series of lines L: $\{l_1, l_2, ..., l_k\}$ with the connecting nodes $\{n_0, n_1, ..., n_k\}$, where n_0, n_k are in V, and $n_1, n_2, ..., n_{k-1}$ are not in V.

Simply speaking, a road network consists of crosses (vertices in G) and road segments (polylines between vertices). While a road segment consists of one or more edges (lines inside a polyline).

To support the efficient path search and the effective region search, RR-tree [70] manages the properties of road segments and detailed spatial information of road networks in an integrated manner. RR-tree consists of three parts: the first part is the ordinary R-tree created by regarding the road segments as basic objects of the road networks; the second part is the storage for the road segments; and the third part is a disk file for storing the detailed spatial information of the road networks. The structures of these three parts are:

- R-tree for road segments: in every entry of internal nodes, there is a record of the left-bottom and right-upper coordinates of MBR, which bounds the spatial contents inside its child node, and a pointer which points to its child node. In the bottom level, MBR corresponds to MBR of a road segment, and the pointer indicates the storage location of that road segment.
- Storage for road segments: for every road segment there is a fixed length record in the storage. The record consists of the ID's of the segment's two end nodes, the length of this segment, the number of edges included in this segment and a pointer which indicates the start location of this segment in Polyline File.
- Polyline File: to store the detailed spatial information of edges which are grouped by segment. In this file, the record consists of the coordinates of the series of segment nodes.

The creation process of this structure is just like creating R-tree for ordinary spatial objects, except for placing a pointer in the storage of segment, which points to Polyline File. Because the structure is created based on the spatial information of the road networks, either the spatial query, such as region query, or the path search can take advantage of this structure [70].

4.3.3 Structure of Sketch RR-Tree

To take the advantage of the efficient management of road networks and solve the distinct counting problem, we integrate RR-tree and sketch method. In other words, we replace R-tree in Tao's structure with RR-tree, and keep the B-trees of sketches for approximate values in the corresponding spatial region. The image of sketch

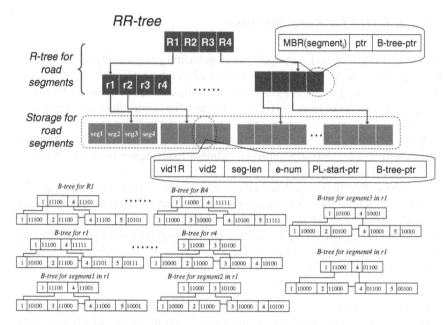

Fig. 4.13 Structure of Sketch RR-tree

RR-tree is depicted in Fig. 4.13: there is an additional attribute (B-tree-ptr) in the record of every node in the RR-tree part, which refers to the corresponding sketch B-tree of every entry region, and also is an additional attribute (B-tree-ptr) in the segment record inside the storage, which refers to the sketch B-tree of every road segment. The query operations of all the leaf nodes and the storage are the same as those in RR-tree and Tao's index structure. Because in RR-tree, the index element is road segment, and the sketch can also be generated for the road segment. Therefore, by using sketch RR-tree, not only the aggregate query on district but also that on segment can be responded efficiently.

4.3.4 Operations on Sketch RR-Tree

The search algorithm of the Sketch RR-tree index structure is similar to an aRB-tree. The search has two parts: RR-tree and B-tree, and starts from the root of the RR-tree. The aggregate query on the moving objects in the segments is given in Algorithm 7.

Similarly, B-tree search also starts from the root. When the level of the B-tree is 0, we only need to check whether the interval of the entries in the level overlaps the query interval. If so, do the OR operation of the corresponding sketch and the result sketch, and store the result into the result sketch array.

When the level of the B-tree is more than 0, i.e., level > 0, we need to check the interval of the entries which overlaps or contains the query interval. If the interval contains the query interval, do the OR operation of the corresponding sketch and the result sketch, and store the result into the sketch array result. If the interval overlaps the query interval, do the aggregate query for the child node of the entries recursively.

After the search of the sketch index structure, we will get the result sketch. Find the position of the first 0 in the array of the result sketch. Let the position be k, using the equation $n = 1.29\,\mathrm{m} \times 2^{k/m}$, we can get the query result.

Algorithm 7: Agg_Qry_RR(sketcharray, segmbr, qt)

---

input : *sketcharray* is used to store the result sketch at present,
segmbr is the query segment,
qt is the query interval.

```
1 begin
2    if level==0 then
3       foreach entry in the level of the intermediate nodes do
4          Check whether the extent of this entry contains the query segment;
5          if contains then
6             Do the aggregate query on the corresponding B-tree;
7             return;
8          end
9       end
10   else if level > 0 then
11      foreach entry in the level of the intermediate nodes do
12         Check whether the extent of this entry contains the query segment;
13         if contains then
14            Invoke the algorithm Agg_Qry_RR for the child node of this entry recursively;
15         end
16      end
17   end
18 end
```

4.3.5 Evaluation

In this section, we evaluate the Sketch RR-tree and analyze the applicability of our method in aggregate query for road network-constrained moving objects. Our experiments used a PC of Pentium IV CPU with 3.19 GHz and 512 MB of main memory. The trees and the algorithms were implemented in C++ and compiled using Visual C++ version 2003. The experimental road network is California road network gathered by Geographic Data Technology Inc in 1999. There are about 1,451 road segments inside per km^2. We use part of the road network (about 806 road segments), and generate 10,000 vehicles and divide them into four kinds (car,

bus, truck and auto-bike) with different speed and moving patterns. The vehicles are
uniformly distributed on the road network at the start time. By recording the positions
(segment-id) of every vehicle for every 5 s, we used FM-PCSA algorithm [68] to
generate sketches and inserted them into our sketch index, and carried out three
kinds of experiments:

- Compare the size of sketches with the data streams. The experiments show that
  when the number of sketches is less than 48, there only need less than 30 % space
  comparing to the original data stream.
- Test the sketch generation time and the aggregate query responding time. The test
  is to evaluate the efficiency of our method. It shows the generating time and the
  query time are all within 100 ms.
- Compare the approximate value responding for the aggregate query with the real
  data. It is to test whether our method can respond the aggregate query within a
  reasonable error or not. We execute the queries on the data sets with different
  numbers of sketches, and compare the average maximum query errors. It shows
  the error is large when using small numbers sketches or large number sketches.

To find out the reasons of the high errors, we did experiments with different
numbers of vehicles. The high errors appeared at the following situations: (1)there
are relative small numbers of vehicles for the number of sketches; (2)there are relative
large number of vehicles for the number of sketches. Therefore, we adopt the middle
numbers of sketches, and compute their average errors. By observing all the average
errors (all are less than 20 %), we can conclude that in a real road network, if the
number of sketches is selected properly, the aggregate query errors can be controlled.
And the number of sketches can be selected by considering the vehicle numbers inside
a region. In the extremity situations, e.g., there are extreme large number of vehicles
at the rush time, we can set the number of sketches as variable.

4.4 DynSketch

4.4.1 Introduction

The sketch RR-Tree index solves the distinct counting problem. However, it is not
suitable for two situations of aggregate queries over moving objects in road networks.
One of the solutions is to adopt the middle number of sketches. However, it is hard
to decide the middle number of sketches in practice. In order to solve the above
problems, we propose a new method "DynSketch" index structure to improve the
quality of queries over network-constrained moving objects in road networks, with
respect to the analysis of the traffic jam in ITS. The experiments show that DynS-
ketch outperforms the sketch method in space consumption, queries efficiency, and
approximation errors control. The DynSketch only consumes small storage space,
has quick response time and efficient query quality, and does well in relatively small
region queries especially.

4.4.2 Histogram

Histogram techniques split the data space into buckets, usually based on the assumption that the data within a small region are almost uniform. The various methods differ on the partition policy, such as equi-depth histogram [72], optimal histogram [73], and compressed histogram [74]. However the data in the above histograms are static, but not suitable for spatio-temporal queries. There are also some multi-dimensional methods, such as dynamic multi-dimensional histogram [75]. In this histogram, the sketch maintains succinctly the stream tuple distribution. Arrivals of new stream tuples are very efficiently reflected on the sketch. A histogram structure of the multi-attribute stream can be derived from the sketch efficiently and on demand. However, the main disadvantage of this approach is that the extraction process is expensive. Therefore, it is not suitable for on line queries. Thus, [69] proposed Adaptive Multi-dimensional Histogram (AMH) suitable for moving objects.

Given a regular $w * w$ cell partition, AMH generates n rectangular buckets whose edges are aligned with the cell boundaries. Each bucket stores: (1) its rectangular extent R, (2) the average frequency f of all the cells in R, and (3) the average of "squared" frequency g of these cells:

$$g_i = (\frac{1}{n_k}) \sum_{\forall c \in R} (F_c)^2 \tag{4.5}$$

Thus, n (the number of cells covered by a bucket) can be represented as $R * w^2$, where w is the cell resolution. A bucket's variance is $v = g - f^2$. As a result, the weighted variance sum (WVS) can be computed as:

$$WVS = \sum_{i=1}^{n} n_i * v_i \tag{4.6}$$

AMH aims at minimizing the WVS of all the buckets.

AMH maintains a binary partition tree (BPT), where each leaf node corresponds to a bucket. An intermediate node is associated with a rectangular extent R that encloses the extents of its children. Initially, BPT contains a single leaf node, and the histogram has one bucket covering the entire data space. New buckets (leaf nodes) are created through bucket splits, but the total number n of buckets never exceeds a system parameter B. As an example, Fig. 4.14 shows the structure of AMH. Figure 4.14a shows the frequencies of 25 cells (i.e., $w = 5$) at timestamp 1, Fig. 4.14b demonstrates the extents of 6 buckets, Fig. 4.14c illustrates the corresponding BPT, and Fig. 4.14d stores every bucket information R, f, g, v.

AMH avoids re-building by using query feedback to refine the buckets. It partitions the area intelligently and has query efficiency with limited memory. However, it cannot solve the distinct counting problem.

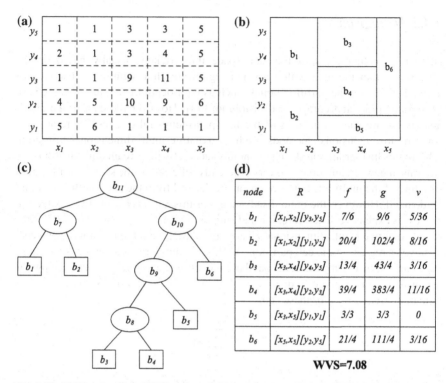

Fig. 4.14 AMH structure. **a** Cell frequency. **b** bucket extents. **c** the BPT. **d** Bucket information

To take the advantage of the sketch index and AMH technology, and solve the problem in these two methods, this article first proposes "fitting sketch", and then develops DynSketch index. In order to improve the quality of the approximation, the DynSketch index intelligently partitions the sketch method by existing histogram technique AMH.

4.4.3 Fitting Sketch

The "fitting sketch" is a method to adjust the number of sketches dynamically. That is to say, when there are relatively small number of vehicles in query region, the number of sketches is decreased, and the vehicles can be aggregated by using small number of sketches; When there are relatively large number of vehicles in query region, the number of sketches is increased, and the vehicles can be aggregated by using a large number of sketches.

The method of curve fitting is used to implement the detailed "fitting sketch". The curve fitting process fits to equations of approximating curves with the raw field data. For a given set of data, the fitting curves of a given type are generally not unique. The best-fit curve can be obtained by the method of least squares.

The method of least-square assumes that the best-fit curve of a given type is the curve that has the minimal sum of the deviations squared (least square error) from a given set of data. Suppose that the data points are $(x_1, y_1), (x_2, y_2), \ldots, (x_n, y_n)$, where x is the independent variable and y is the dependent variable. The fitting curve $f(x)$ has the deviation (error) d from each data point, i.e., $d_1 = y_1 - f(x_1)$, $d_2 = y_2 - f(x_2), \ldots, d_n = y_n - f(x_n)$. According to the method of least-square, the best fitting curve has the property:

$$d_1^2 + d_2^2 + d_3^2 + \cdots + d_n^2 = V_{min} \tag{4.7}$$

where V_{min} is a minimum value.

In this method, first, the number of the sketches used in sketch index is set as x, the corresponding actual value of moving objects are as y, and the corresponding approximate value is set as est (i.e., $f(x)$). The approximate value est is genereted by sketch index method when choosing different number of sketches (i.e., x). Then, three typical fitting curves are choosen:

$$y = b * x + a \tag{4.8}$$

$$y = a * x^b \tag{4.9}$$

$$y = a * e^{b*x} \tag{4.10}$$

According to the diffenent (x_i, y_i) pairs, the best fitting curve will be found by the method of least-square. However, how to set (x_i, y_i) pairs, and which is the best fitting curve for moving objects in road networks?

V_{min} is set as 0.001. For one value of y (e.g., the number of moving objects is 500, 1,000 or 5,000), x varies from 1 to 96 (i.e., the minimum and maximum numbers of sketches used in sketch index for the moving objects in road networks). By the method of least-square, the correlation coefficients of the equations are computed, and it is found that the best fitting curve for moving objects in road networks is:

$$y = 0.747978 * x^{0.708563} \tag{4.11}$$

Through this fitting sketches method, we can adopt fitting number of sketches by the number of moving objects.

4.4.4 Framework

DynSketch index is a dynamic sketch index structure for aggregate queries over the network-constrained moving objects in road networks. In DynSketch index, the numbers of sketches are generated by "fitting sketch" method.

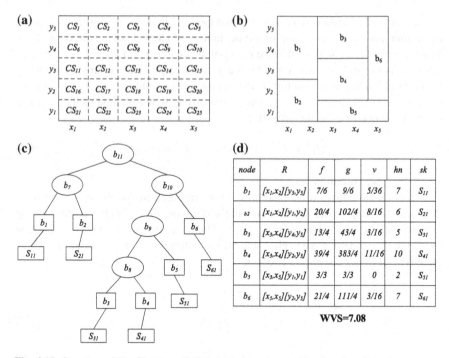

Fig. 4.15 Structure of DynSketch. **a** Cell frequency represented by sketches. **b** bucket extents. **c** BPT sketch information. **d** Bucket information added sketch information

Figure 4.15 shows an example of the structure of DynSketch.

This is similar to the AMH structure in Fig. 4.14. However, there are also some differences as follows. It consists of two parts: sketch and histogram.

(1) In Fig. 4.14a, the actual number of moving objects in every cell represents cell frequency. However, in DynSketch index structure (in Fig. 4.15a), sketch information of the moving objects in every cell represents cell frequency. For example, CS_i in Fig. 4.15a represents the sketch information of the moving objects in the i-th road segment. The advantage of DynSketch is that it solves the distinct counting problem in every cell, and decreases the space consumption.

(2) In Fig. 4.14c, each leaf node of BPT is a bucket. However, in DynSketch index structure (in Fig. 4.15c), each leaf node of BPT is sketch information. For example, S_{bt} in every leaf node represents sketch information of bucket b at time t. The access of BPT in DynSketch is depicted in detail in Sect. 3.4.

3) In DynSketch index structure (in Fig. 4.15d), we add sk (sketch information of buckets) and hn (the number of sketches for the corresponding buckets) into bucket information. The advantage of Dynsketch is that it solves the distinct counting problem in every bucket.

4) In DynSketch index structure, the update algorithm of buckets and road segments, and the query algorithm are completely different from those in AMH structure. We will describe it in the following.

4.4.5 Update of Buckets and Road Segments

The update algorithm of buckets and road segments in DynSketch is Algorithm 8: $Info\_Update()$.

Algorithm 8: Info_Update(*tm*, *rid*, *eid*)

---

input : *tm* is the time when the data is sent, *rid* is the road id where the vehicle is at *tm*, *eid* is the vehicle id

1 **begin**
2   Find the cell c (the vehicle be in) by a hash function;
3   Descend BPT to find the bucket b that contains the cell c;
4   FM_PCSA($eid, hahnum, CS_i$);   /* update sketch information of the cell CS_i */
5   FM_PCSA($eid, hahnum, S_{bt}$);   /* update sketch information of the bucket S_{bt} */
6 **end**

When a new vehicle is coming, first, traffic supervision system gets the data $<tm,$ $rid, eid>$. For example, at time 1, a vehicle at the 8-th road segment sent data $<1, 8, 2389>$ to the control center. Second, according to a hash function, algorithm $Info\_Update()$ finds a cell c where the road segment will be in. The hash function is that: as the integer of 8/5 (i.e., rid/w) is 1, and the remains of 8/5 is 3, the 8-th road segment is in the 1st row, and 3rd column is in the array (i.e., $[x_3, y_1]$ in Fig. 4.15a. Then, $Info\_Update()$ traverses BPT to find the bucket b which contains c. Finally, it adopts fitting number of sketches (generated by "fitting sketch" method) to generate sketch information of the road segment and the corresponding bucket. After processing all objects, the information of road segments and the corresponding buckets are updated.

4.4.6 Algorithm of Search Using DynSketch

The search algorithm of the DynSketch index is similar to that of the AMH structure. However, the difference is that the average frequency of the buckets in the DynSketch index is computed by sketches. In addition, it considers the distinct counting problem in DynSketch. The procedure of search using DynSketch is Algorithm 9.

First, Algorithm 9 initializes all bits of sketch array as 0 (line 2). When there is a query region (line 3), if the BPT is not empty (line 3), Algorithm 9 pre-order traverses the BPT, in order to find out the buckets which have been overlapped with the query region (line 5). Then, for every bucket b_i which has been found, Algorithm 9 computes the overlaped area R_i (line 7), and does OR operation on the sketches at every timestamp in bucket b_i by algorithm FM_PCSA (lines 7, 8). After that, Algorithm 9 gets the total number of vehicles *allnumber* in buckets which had been traversed (line 12), and all the area of traversed buckets n (line 13). As a result, the number of vehicles in query region is $f * R$, where average frequency $f = allnumber/n$.

Algorithm 9: Info_Search(*root,query_region,query_interval,allnumber,qbitmap,n*)

input : *root* is the root of BPT,
   query_region is the query region,
   query_region is the query interval,
   allnumber is all the number of the vehicles until now,
   qbitmap is sketch arrary,
   n is the sum of area covered by all the buckets.

1 **begin**
2 | Initialize all bits of sketch array *qbitmap* as 0, *allnumber=0*;
3 | **while** *here is a query region* **do**
4 | | **if** *BPT is not empty* **then**
5 | | | Pre-order traverse BPT to find out buckets b_i overlaped with query region;
6 | | **end**
7 | | **foreach** *bucket overlaped with query region* **do**
8 | | | Compute the area R_i overlaped by query region and the bucket b_i;
9 | | | Do OR operation on sketches in b_i at every timestamp with *qbitmaps* estimate the approximate value of vehicles in b_i by FM_PCSA: *est* get all number of vehilces until now: *allnumber=allnumber+est*;
10 | | **end**
11 | **end**
12 | Compute all the area of traversed buckets: n, and all overlaped area R;
13 | Compute the average frequency f by *allnumber/n*;
14 | Compute the number of vehicles in query region $f * R$;
15 **end**

Take the structure of DynSketch in Fig. 4.15. For example: the query interval is [1, 2], and the query region is in Fig. 4.16. Suppose that the areas of the buckets are R_{b_1}, R_{b_3}, R_{b_4} respectively. Algorithm $Info\_Search$ finds the buckets b_1, b_3, b_4 by pre-order traversing the BPT. Then, it computes the overlap areas in corresponding buckets, and marks them as R_1, R_3, R_4. After that, it does OR operation on the sketches in the buckets at time 1 and time 2, to get the sketches S_1, S_3, S_4, and the number of vehicles n_1, n_3, n_4 by Eq. (4.12). At the same time, it computes the average frequency of the buckets f_1, f_3, f_4. In the end, it does OR operation on S_1, S_3, S_4 to get the final sketch S. N is the distinct counting in all the buckets. Thus the query result is y:

$$y = f_1*R_1 + f_3*R_3 + f_4*R_4 - [(n_1+n_3+n_4-N)/(R_{b_1}+R_{b_3}+R_{b_4})]*(R_1+R_3+R_4)$$
(4.12)

where $(n_1 + n_3 + n_4 - N)/(R_{b_1} + R_{b_3} + R_{b_4})$ is the number of the average distinct counting.

Fig. 4.16 Query region QR

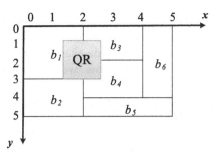

4.4.7 Evaluation

4.4.7.1 Experimental Settings

The experiments used a PC of Pentium IV CPU with 3.19 GHz and 512 MB of main memory. The trees and the algorithms were implemented in C++ and compiled using Visual C++ version 2003. The experimental road network is California road network gathered by Geographic Data Technology Inc. in 1999. There are about 1,451 road segments per km². We use part of the road network (about 784 road segments), generate 10,000 vehicles and divide them into four kinds (car, bus, truck and auto-bike) with different speed and moving patterns. The vehicles are uniformly distributed on the road network at the start time. We record the positions (segment-id) of every vehicle for every 5 s.

The experiment has three steps mainly. First, according to the different history number of vehicles in every road segment, all the road segments are split into different numbers of histogram buckets varied from 50 to 784. Then dynamic number of sketches are used to record the information of moving objects inside every bucket. The number of sketches is decided by "fitting sketch" method which we have introduced. Finally, queries are executed on the DynSketch structure based on the histogram.

4.4.7.2 Experimental Results

In the sketch index structure, the efficiency of using 48 sketches to aggregate queries over the network-constrained moving objects is best. Thus, in order to compare easily, the method DynSketch is compared with 48 sketches. Three kinds of experiments are carried out to compare DynSketch index with sketch index in three aspects: space consumption, queries efficiency and approximate errors. The three experiments are as follows:

(1) Compare the space consumption of DynSketch with that of 48 sketches. With the increase of the number of the buckets, the increase of the space consumption is slow. Thus, we can get the result that the DynSketch index is better than the sketch method on the space consumption.

(2) Compare the generation time and the query response time of DynSketch with those of the sketch technique. Although the generation time of the DynSketch is more than that of the sketch method, the generation time of the DynSketch is still within 100 ms. Thus, we can get the result that the efficiency of the DynSketch index in queries outperforms that of sketch technique.

(3) Compare the approximate value responding for the queries in the DynSketch method with that in 48 sketches. It is executed to test whether the DynSketch can provide more accurate answers compared with the sketch method. The experiment is carried out by using three series of random queries. There are also three experiments which compare the maximum errors, the average errors and the boundary errors in DynSketch method with those in 48 sketches. The experiments are as follows.

(1) The first experiment is carried out to compare the maximum error in DynSketch method with that in 48 sketches. We can see that the maximum error is significantly decreased in DynSketch index.

(2) The second experiment is carried out to compare the average query errors in the DynSketch with that in 48 sketches by executing the three series of queries. We can get the result that for the relatively small region queries, DynSketch index has significantly predominance.

(3) The third experiment is carried out to do the query in the boundary condition of the DynSketch method. The results show that the DynSketch can get the approximate result effectively, especially for the relatively small region queries.

4.5 Modified Histogram

4.5.1 Introduction

Traditional methods mainly focus on accessing precise position data of individual objects, in history [67, 76], and currency [76, 77], or both [78]. However, recently, many applications (e.g., ITS, network, large telecommunication) require summarized spatio-temporal data, rather than information about the locations of individual objects in time. There are only limited memory space, and the actual number of individual objects may be highly volatile and need extreme space, while the aggregated data may remain for long intervals, because this requires considerably less space for storage. In a sort of sense, the aggregate information is more useful than the precise position data of individual objects.

Therefore, different modern ITS systems start to pay attention to the aggregate information about traffic data streams (e.g., the number of vehicles) to analyze different problems like the traffic jam: e.g., a natural question in traffic analysis is to ask how many vehicles stay in a query region during a query interval before doing further computation. And how to get the aggregate information over the traffic data streams in road networks efficiently starts to be a new challenge to the moving objects database community.

4.5.2 Motivation

In conventional databases and stream management systems, approximate query based on aggregate information has been addressed using various techniques such as sketches [79, 80], histograms [74], sampling [81], wavelets [82]. However, all these methods are not effective in supporting spatio-temporal query.

In spatio-temporal databases, there are two problems which are difficult to deal with. The first problem we must face is the distinct counting problem: if an object remains in the same query region over several time intervals, it will be counted multiple times as the result. An effective solution to distinct counting is vital for several applications because it enables much richer range of decision-making queries. For instance, asking how many vehicles stayed in a region is a natural question in traffic analysis. The former techniques such as [67] can be used to find the average number of vehicles per timestamp during a time interval. However, the average is not sufficient in analyzing the traffic volume. For example, one might find a parking lot and a busy highway with the same average number of vehicles, and then assert that they have similar traffic characteristics. If we introduce distinct counting queries, we can easily distinguish between those two cases. First, the highway will have much higher turnover rates of entering and leaving vehicles which will be directly reflected in a higher number of distinct vehicles over time. Second, this turnover rate can be quantified by comparing this number against the average. That is, if the number of distinct vehicles increases by x vehicles, while the average stays constant, x must have entered and x must have left within the time period in question. The difficulty for supporting distinct counting queries is due to the fact that there is no way to exactly summarize distinct objects substantially better than simply enumerating all of them.

The second problem is that it may cause big error and high space consumption which exist in the sketch method. Here we take Adaptive Multi-dimensional Histogram (AMH) [67] into consideration to deal with this problem: AMH divides the space and completes the inquiry fast in the limited memory space. However, AMH has the following shortcomings: first, it cannot handle the distinct counting problem; second, although bucket's number is critical for the quality of answers, no value of bucket's number is to ensure the quality for all queries; third, the reorganization of cells partition is only performed when the system is free, so that the quality of the answer continues to deteriorate during consecutive reorganization operations. In consideration of AMH's last two shortcomings, we use Adaptive Multi-dimensional histogram* (AMH*) [69] which is an improved version of AMH structure for reference; AMH* splits the whole area into lots of buckets, but the number of buckets can grow or shrink according to the change of the data distribution, it can find a balance between bucket's number and inquiry quality effectively.

To avoid these two problems effectively, in this section, a new method which effectively takes the advantage of existing approximate statistical information on the stream is proposed to decompose the sketch problem in a way that provably enhances the estimation guarantees.

4.5.3 Framework

Modified Histogram is a novel method for aggregate queries over the network-constrained moving objects in road networks. We assume a set of objects (vehicles) move in the two-dimensional space. Moving objects may issue updates in periodic intervals or based on the deviation from the previous transmitted position. Here we adopt a 2-D grid that partitions the data space into $w * w$ (where w is a constant called the *resolution*) regular cells with width $\frac{1}{w}$ on each axis. Each cell c ($1 \leq c \leq w^2$) is associated with a *frequency* F_c, which is the number of objects in its area. MH combines adjacent cells with similar frequencies into a small number of (rectangular) buckets. For each bucket b_k ($1 \leq k \leq n$), we denote: (1) n_k as the number of cells it covers, (2) f_k as the average frequency of these cells (i.e., $f_k = (\frac{1}{n_k}) \sum_{(\forall \, cell \, in \, b_k)} F_c$), and (3) v_k as their variance (i.e., $v_k = (\frac{1}{n_k}) \sum_{(\forall \, cell \, in \, b_k)} (F_c - f_k)^2$). One object of MH's index structure is to minimize the *WVS* in all the buckets, that is $WVS = \sum_{i=1-n} (n_i * v_i)$.

The structure of MH is shown in Fig. 4.17. Each bucket stores: (1) its rectangular extent R, (2) the average frequency f of all the cells in R, (3) the average of "squared" frequency g of these cells: $g = (\frac{1}{n_k}) \sum_{(\forall \, cell \, in \, R)} (F_c)^2$, (4) sketch number hn, and

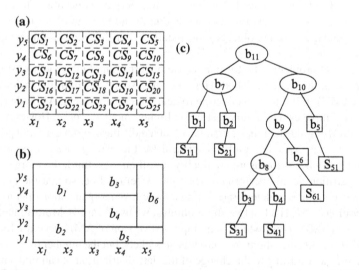

Fig. 4.17 Structure of modified histogram. **a** Cell frequency. **b** bucket extents. **c** the BPT

(5) specific sketch information s_k. Thus, the number n of cells covered by a bucket can be represented as $R * w^2$, where w is the cell resolution. A bucket's variance v equals $v = g - f^2$, and as a result, WVS can be computed using R, f, g of all buckets.

MH maintains a binary partition tree (BPT), where each leaf node corresponds to a bucket. An intermediate node is associated with a rectangular extent R that encloses the extents of its (two) children. Initially, BPT contains a single leaf node, and the histogram has one bucket covering the entire data space. New buckets (leaf nodes) are created through *bucket splits*. As an example, Fig. 4.17a shows the frequencies (it is computed by sketch information) of 25 cells (i.e., $w = 5$), Fig. 4.17b demonstrates the extents of 6 buckets, and Fig. 4.17c illustrates the corresponding BPT. Intermediate node b_8, for example, covers buckets b_3, b_4, implying that they were created by splitting b_8.

4.5.4 Evaluation

The Sketch [79, 80] and DynSketch [83] are evaluated for sketch generation, the query time, the need of main memory, and the approximation error of aggregated queries. As to explain the superiority of our method, we make an overall analysis from the following several aspects: First, through transforming the query range, the different experiment results are compared to obtain a suitable query range of the method. Then, our method is compared with Sketch and DynSketch to verify, and is more efficient.

4.5.4.1 Analysis of Suitable Query Range

The same parameters which are in [79]: $\varepsilon = 0.5$, $\rho = 0.95$ ($\eta(\rho) = 1.96$) are adopted. (1) Find suitable query region interval. Take the different query region interval and the same time interval(here we set 5), generate the query region data of vehicles' information on the road segment. In each query region interval, generate average relative error and maximum relative error. When region interval is set inside 3–5, both kinds of errors are very small. (2) Find suitable query time interval. Take different query time intervals and the same region interval (the value is 3 which is got from foregoing experiment), randomly generate the query time data of vehicles' information on the road. And both two kinds of errors are smallest when time interval is set inside 5–10.

4.5.4.2 Compare with Sketch

In the sketch index structure, from some history experiment result, we know that the efficiency of using 48 sketches for aggregate queries over the network-constrained moving objects is the best. We carried out four kinds of experiments:

(1) Compare the sketch generation time. Along with the increment of the number of sketches, sketch generation time gradually increases, and it needs more time. MH's generation time is effectively controlled in a range of reasonableness, and the phenomenon of the deterioration does not occur.
(2) Compare the query time, the aggregated query time has markedly decreased nearly 24 times. Average generated query time is less than 0.1 ms and many of them are close to 0 ms. So, our method has faster query efficiency.
(3) Compare the storage space. The account of storage space required is 20 stream. Therefore, for the space requirement, compared with sketch, our method has greatly improved.
(4) Compare the query error. Although comparing average relative error with 48 sketches, the method is 1–2 % larger, this method is still effectively controls the error between 11–12 %.

4.6 Summary

Index techniques are used to improve the efficiency of query. Road network data is a typical example of spatial data, and index mentioned above such as R-TPR tree, MOR-tree, DynSketch, Modified Histogram and CR-tree can meet the requirements and improve query efficiency.

Especially, R-TPR tree is a composite index structure for managing the real-time data of moving objects on the road network. MOR-tree is created for multi-levels of road networks under M^2 map information model and R-trees are for the datasets in every level managed by L-model, respectively. DynSketch uses AMH to partition Sketch RR-tree intelligently to deal with the distinct counting problem and non-uniform distribution of moving objects. Modified Histogram (MH) is an index technology to partition the space automatically and improve the quality of the approximation.

Index does not sustain all needs in road network as there are many query requests in road network. In the next chapter we will center on the typical query methods in road networks.

Chapter 5
Query in Road Network

There are two typical kinds of queries in road network: precise query and aggregate query. Nearest neighbor query (NN) and continuous nearest neighbor query (CNN) belong to precise queries which would get exact location in road network and are used widely in ITS. The non-Euclidean property of road network is the most significant problem in these queries. Cyclic Optimal Multi-step Algorithm divides the temporal query into two stages: filter and refinement. The algorithm uses non-Euclidean semantic distance as filter condition to generate candidates. Followed by refinement step, the filter conditions are constantly revised and the above process is iterated to enhance the retrieval efficiency by using the spatial index created in Euclidean space. Sections 5.1–5.3 introduce nearest neighbor query, continuous nearest neighbor query and reverse search method of CNN based on Cyclic Optimal Multi-step Algorithm.

Aggregate query aims at obtaining summarized information such as vehicles' count. In this kind of query, distinct counting problem and non-uniform distribution problem are more prominent. For example, when we want to know how many vehicles stay in a query region within a query interval, some vehicles may be computed multiple times in the period of time, which is very inefficient. What is more, vehicle density of some road segments is larger than that of other road segments, then it is difficult to get query results at the same time unless using suitable methods. Forecasting methods can support decision made by managers in transportation field. Section 5.4 introduces Sketch-based methods such as ES (Exponential Smoothing forecasting method), The Self-Adaptive Exponential Smoothing (SAES) and The Smooth Transition Exponential Smoothing (STES) methods are explained to solve the above problems in aggregate query.

© Springer International Publishing Switzerland 2015   107
J. Feng and T. Watanabe, *Index and Query Methods in Road Networks*,
Smart Innovation, Systems and Technologies 29,
DOI 10.1007/978-3-319-10789-9_5

5.1 Nearest Neighbor Search on Road Network

5.1.1 Introduction

Most use of GIS multiple layers provides different aspects for the modeled real world. Geographic objects in every layer are indexed by specific spatial structures (e.g., R-tree). In order to solve queries relating to multiple layers, a layer-overlay operation has to be performed first. Then, the search may take advantage of the spatial index, and start a filter-refinement step so as to respond to the query. This search method becomes very computational-intensive when there are a large number of objects in referred layers.

Therefore, this section gives an advanced method, called Cyclic Optimal Multi-Step (COMS) method, to search the nearest target object on the existing road network without overlaying the target object and the road network layers. It centers on the bridging between straight-line-based filtering step, operating on the dataset of target objects, and path-length-based refinement step on the dataset of road network. Analysis and evaluation in the end of this section show a good performance of this method.

5.1.2 Framework of Cyclic Optimal Multi-step Method

To search spatial datasets of multiple layers, there is a multi-step (denoted as MS) spatial query method. MS method takes advantage of spatial structures which index the spatial datasets, and consists of two sequence steps of filtering and refinement [84]. MS method generates all candidates (a set of candidates) based on specific filtering conditions from the underlying index structure, and then enters refinement step to get results by testing the exact representation of every candidate. The two steps run in a sequence for one time. Because the refinement step runs on real spatial data set, the test is usually time consuming, especially when there are too many candidates.

To solve this problem, a COMS method which consists of two interactive steps running in cycles was proposed. COMS method generates only one candidate in the filtering step for one time, and then tests this candidate in the refinement step. If the candidate cannot be the result, COMS method generates new filtering conditions based on the refinement step of this time and re-runs filtering step to generate the next candidate. As only one candidate is generated at the filtering step for one time, the filtering conditions are much more strict than those in MS method; moreover, the two steps run cyclically, and the conditions can be adjusted to a better one every time. COMS method is depicted in Fig. 5.1.

To search nearest target object based on the road network, there are at least two datasets: the road network and the target object datasets. It is assumed that the two datasets are indexed by spatial structures (R-trees), respectively. By using

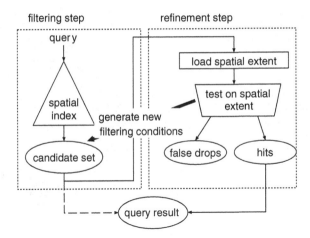

Fig. 5.1 Cyclic optimal multi-step spatial query processing. Filtering step is based on index structure of target objects and refinement step is based on road network

MS method, the query is answered by overlaying the two datasets first. When the two datasets are large, this operation becomes very computation-intensive. Here, we give ideas for nearest neighbor search by using COMS method:

(1) Although the straight-line distance is not completely consistent to the path length in the road network, the object with shorter distance from the source has a higher possibility to get shorter path length; so, by using the R-tree structure of target objects, the object is selected in the order of straight-line distance as the candidate for computing path length.

(2) The candidate cannot be regarded as the result, but the path length from the source object to this candidate, computed on the basis of the road network, can be regarded as a limit for the following search. In other words, only the elements (objects and R-tree nodes), whose straight-line distance from the source is shorter than the path length of the candidate, need to be re-checked.

The practical viewpoint for these ideas is based on a condition relating to the data structure of R-tree and a proposition bridging between the filtering step and refinement step.

Condition 5.1:

The distance in a straight-line between a source object s and a node N in R-tree is no longer than the distances between s and any elements or child node e inside the node:

$$D(s, N) \leq D(s, e) \tag{5.1}$$

Here, $D(x, y)$ represents the minimum straight-line distance between MBRs of objects or R-tree nodes.

Fig. 5.2 The *circle* around
source object *s* depicts the
region after calculating the
path length from *s* to *t* as the
up-to-now minimum path
length. For simplicity, leaf
nodes are represented by
grids. In the following search,
only the objects inside *shaded*
leaf nodes are re-checked, and
the computing of shortest
path is based on the roads
inside the same region

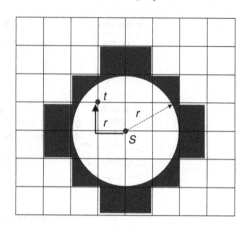

This condition is strictly tied to the hierarchy defined by the data structure—
R-tree. Each node in R-tree stores a maximum of n entries. Each entry consists of a
rectangle *Rec* and a pointer *Pt*. For nodes at the leaf level, *Rec* is the bounding box
of an actual object pointed by *Pt*. At internal nodes, *Rec* is the minimum bounding
rectangle (MBR) of all rectangles managed in the subtree pointed by *Pt*. Under this
condition, we limit the target objects, which need to be located on the road network,
to the elements with shortest straight-line distance. Let t be the target object, s be the
source object, and r be the path length from s to t.

Proposition 5.1 *If exist the objects which are nearer to s than up-to-now nearest
object t,they can be found within a circle C (the shaded circle region in Fig. 5.2). The
radius r of C is the nearest path between s and t.*

This proposition is a bridge between the filtering and refinement steps: at the
filtering step a candidate is selected by taking advantage of previous condition; at the
refinement step the path length based on the road network is computed and just this
path length is used to limit the following filtering step. Consider that the distance of
a straight-line between a nearer object and the source object is not longer than that
of the up-to-now shortest path: in other words, only the objects, whose straight-line
distance is shorter than that of the up-to-now shortest path, need to be considered in
the following process; and at the same time, if the straight-line distances from the
source object to both ends of a road segment are longer than that of the shortest path,
this road segment can only lead to a longer path length. So, the road set can also be
limited by the up-to-now shortest path. The region within r from s is called a search
region. We assume that s is a point and the search region is a circle with radius r.
Figure 5.2 depicts this conceptual view. The value of r is decreased by keeping step
with the ongoing path length computation for the candidates and the search region
is adjusted with new r. The stepwise refinement will not be ended until there is no
candidate inside the search region.

5.1.3 Cyclic Optimal Multi-step Algorithm

The algorithm of our COMS search process bases on a priority queue and R-trees for objects and road network datasets. The priority queue is used to record the intermediary results in the order of length: R-tree nodes which are overlapped with the search region and the target objects are necessary to compute the shortest path from a source object. The key by which the elements on the queue are to be sorted is straight-line distance computed for R-tree node or path length computed for target object. COMS search algorithm is given as Algorithm 1.

Algorithm 1: COMS-search

    input : road network, a set T of target objects; source object *S*.
    output: the nearest object *n*.

1 **begin**
2   Initialize a priority queue *Queue* and radius of search region *r*;
3   Set search-based road-set as the road segments inside the search region;
4   Locate the first node overlapping with the region on R-tree, and insert it into *Queue*;
5   **while** *Queue* ≠ *NULL* **do**
6   **if** *the head of Queue is a leaf node of R-tree* **then**
7   Search the shortest path from *S* to the object inside the head element based on road network;
8   Insert this object to *Queue* in order;
9   **if** *the path length is shorter than r* **then**
10   Reset *r* with this path length;
11   **end**
12   **else**
13   Compute straight-line distances between the source point and the nodes inside the head element;
14   Insert these nodes to *Queue* in order;
15   **end**
16   **if** *the head of Queue is an object t with computed shortest path* **then**
17   Return *t* as the nearest object *n*;
18   **end**
19   Remove *the head of Queue* ;
20   **end**
21 **end**

The priority queue *Queue* is initialized as a node of the target object R-tree, and the search region is initialized as a Maximum. When a target object turns out on the head of the priority queue, it becomes the candidate for further computation. It is located to the road network, and the path length is also computed for the candidate based on the search region. If the path length is shorter than the key of the head element of *Queue*, the candidate is a result. Otherwise, when the path length is smaller than the radius of search region, the search region is reset with the new length as a radius. The value of radius is decreased by keeping step with the ongoing path length computation for the candidates and the search region is adjusted with the current shortest path length

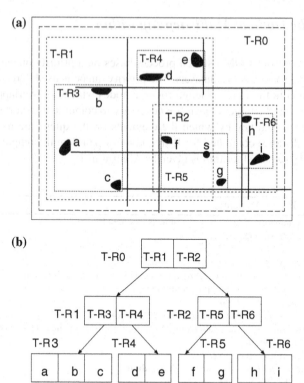

Fig. 5.3 Spatial rendering of target objects. **a** roads and minimum bounding rectangles for target objects. *Dotted boxes* represent nodes in R-tree (*T-R*) and *solid lines* represent roads. **b** *T-R:* R-tree access structure for target objects

until there is no candidate inside the search region—the head element of *Queue* is an object with computed path length.

As an example, suppose that we want to find the target object nearest to source object *s* from all the target objects indexed by R-tree *T-R*, given in Fig. 5.3, where the objects (in the interest of clarity, the bounding rectangles for individual objects are omitted from figures) are polygons stored externally into R-tree, and the road segments (represented by solid lines) stored externally into another R-tree.

Table 5.1 shows the states at every loop of COMS search algorithm. The first column in the table depicts the loop number with the corresponding radius *r* in that loop. The following columns depict the contents of *Queue* in every loop. The contents consist of two parts. One is R-tree node or object name: a node is denoted by its name given in Fig. 5.3 and bounding rectangles of objects are denoted by the corresponding object names embedded in brackets in the table. Another is the corresponding straight-line distance or path length computed for them. An initial search region is r_{max}. The algorithm starts by en-queuing the root node T-R0 of *T-R* into *Queue*, and then enters into the following loops: loops 1–4 are in filtering steps,

Table 5.1 Contents of *Queue* at execution loops of finding nearest object for a source point

	1	2	3	4	5	6
Loop 1(r_{max})	T-R1(0)	T-R2(0)				
Loop 2(r_{max})	T-R2(0)	T-R4(60)	T-R3(80)			
Loop 3(r_{max})	T-R5(0)	T-R6(30)	T-R4(60)	T-R3(80)		
Loop 4(r_{max})	[g](20)	[f](25)	T-R6(30)	T-R4(60)	T-R3(80)	
Loop 5(*100*)	[f](25)	T-R6(30)	T-R4(60)	T-R3(80)	g(100)	
Loop 6(*70*)	T-R6(30)	T-R4(60)	f(70)	T-R3(80)	g(100)	
Loop 7(*70*)	[i](40)	[h](50)	T-R4(60)	f(70)	T-R3(80)	g(100)
Loop 8(*50*)	i(50)	[h](50)	T-R4(60)	f(70)	T-R3(80)	g(100)

R-tree nodes in *Queue* are sorted based on the straight-line distance between *s* and the nodes. The first object with shortest straight-line distance from *s* is brought to the refinement step and located to the road network as a candidate object; in loop 5, the shortest path from *s* to this candidate is computed based on the road network, and the radius of search region is also reset as the new smaller one. Here, r_{max} is regarded as a greater value than 100, and the new radius is set to 100. In the next loop, the filtering steps are executed inside the search region, only the element before the previous candidate in *Queue* needs to be rechecked. The value of radius *r* is decreased repeatedly by the ongoing loops (loops 6 and 8). The algorithm stops after loop 8 by finding out that there is no more candidate before an old candidate object *i* in *Queue*. The object *i* is the nearest object for *s* based on the road network.

5.1.4 Algorithm for Theoretical Analysis

In order to evaluate the algorithm, it is compared with MS method, which does spatial join to locate all target objects on the road network and then uses Dijkstra's algorithm (for an easy comparison we also use this method for computing shortest path from the source object to a candidate in COMS method) to find the nearest target object.

We assume both road segments and target objects are indexed by R-trees respectively: N_r is the number of road segments in the road network; N_n is the number of nodes (cross-nodes among road segments or end-nodes of road segments) in the road network; and N_t is the number of target objects.

The cost of MS method consists of two parts. One part is the cost for overlaying the target objects and the road network. Owing to the property of R-tree [17], the search comparison test is "overlapping": more than one subtree under a node may need to be visited during a search; so, the worst performance is not logarithmic but linear order in the number of keys. The cost of locating N_t target objects on N_r road segments would be $O(N_t \times N_r)$, which means that the computing time is directly proportional to N_t and N_r. The larger the number of target objects, road segments or both of them is, the higher the cost is. Another part is the cost of Dijkstra's algorithm for searching shortest path, based on the overlaid dataset. The complexity is $O(N_r + N_n log N_n)$, where N_r and N_n are the number of road segments and the number of nodes in the road network, respectively.

To derive performance measurements from our method, we analyze the situation after calculating the path length from the source object to a candidate, and the path length is just regarded as a new radius of search region. Furthermore, before proceeding we point out that the algorithm does not locate any target objects on the road network outside the search region and the shortest path search is also based on the road segments inside the search region. This follows directly from the queue order, the condition and the proposition in Sect. 5.1.2.

The measurement is based on the approach, proposed by Henrich [85]. It assumes N_t objects are uniformly distributed as data points in two-dimensional interval $[0, 1] \times [0, 1]$. Also, the leaf nodes are assumed to form a grid at the lowest level of the spatial index with average occupancy of c points, and the search region is assumed to be completely contained in the data space. Assume that points are uniformly distributed, the expected area of search region for the first selected target object is $1/N_t$ and the expected area of the leaf node regions is c/N_t. The area of a circle of radius r_d (the radius drived from straight-line distance) is πr_d^2: so, for the search region we have $\pi r_d^2 = 1/N_t$, which means that its radius is $r_d = \sqrt{1/(\pi N_t)}$. The leaf node regions are squares: so, their length is $s = \sqrt{c/N_t}$. If the path length r_p is k times as long as the straight-line distance, then $r_p = kr_d = k\sqrt{1/(\pi N_t)}$. Henrich [85] points out that the number of leaf node regions intersected by the boundary of the search region is the same as that intersected by the boundary of its circumscribed square. Each of the four sides of the circumscribed square intersects $\lfloor 2r_p/s \rfloor < 2r_p/s$ leaf node regions. Since two adjacent sides intersect the same leaf node region at a corner of the square, the expected number of leaf node regions intersected by the search region is bounded by

$$4(2r_p/s - 1) = 4\left(2kr_d/s - 1\right) = 4\left(\frac{2k}{\sqrt{\pi c} - 1}\right) \tag{5.2}$$

It is reasonable to assume that, on the average, half of c objects in these leaf nodes are inside the search region, while the other half are outside. Thus, the expected number of objects remained in the priority queue (the points in the dark shaded region in Fig. 5.2) is at most

$$\frac{c}{2}4\left(\frac{2k}{\sqrt{\pi c} - 1}\right) = \frac{4}{\sqrt{\pi}}k\sqrt{c} - 2c \approx 2.26k\sqrt{c} - 2c \tag{5.3}$$

The number of points inside the search region (the circle region in Fig. 5.2) is

$$\pi r_p^2 N_t = \pi\left(\frac{k^2}{\pi N_t}\right)N_t = k^2 \tag{5.4}$$

Thus, the expected number of points in leaf nodes intersected by the search region is at most $k^2 + 2.26k\sqrt{c} - 2c$. Since each leaf node contains c points, the expected number of leaf nodes that are accessed to get these points is bounded by $k^2/c + 2.26k/\sqrt{c} - 2$. This means the worst situation: the first selected candidate for path search is just the object with the shortest path length, and all the objects inside the search region generated by the shortest path have to be checked. However, since they give no help to decrease the radius of search region, the expected number of leaf node accesses is $O(k^2)$. The number of road segments inside the region can be computed as the same as that of the target objects. The expected number of road segments is

$$\left(k\sqrt{\frac{N_r}{N_t}}\right)^2 + 2.26k\sqrt{\frac{N_r}{N_t}}\sqrt{c} - 2c \qquad (5.5)$$

The cost of locating these objects on road network inside the search region is $O(k^4 N_r/N_t)$, and the cost of using Dijkstra algorithm inside the search region is $O(k^2)$, which means that the cost is not related to the number of target objects and the complexity of road network directly but is related to the ratio of path length to straight-line distance of the candidate.

5.1.5 Evaluation

A set of experiments are presented to evaluate the COMS method. There are two kinds of networks:

Grid: The synthetic grid graphs correspond to grid patterns with target object distributed uniformly or randomly. Grid graphs are used in experiments, since the parameters of them are easy to control, such as the number of road segments, the number of target objects and their distributions.

Real map: We had the real road map of Ichinomiya city in Japan and used some sub-maps of this real map to verify the validity of the synthetic grid graphs. The used sub-maps are in fixed size $(2,000\,\mathrm{m} \times 1,500\,\mathrm{m})$ with different numbers of road segments (N_r). In our tests, the number of road segments (N_r) is from 358 to 1,642, and the number of target objects (N_t) is set as 20 and 40.

Dijkstra's algorithm is used for path search and experiments were done on an SGI O2 R5000 SC 180 entry-level desktop workstation. Every experiment was repeated 6 times and the results presented here corresponding to an average over these 6 runs. 6 runs were based on 6 different source points. Result variations between 6 runs were caused by (1) the differences among the number of tested candidates; and (2) the differences among the length of computed shortest paths.

We can make some observations as follows:

(1) By using the COMS method, we compute the nearest target object which begins from selecting candidate object based on the straight-line distance between the

source point and target object, and then compute the path length (real distance based on road network) from the source point to the candidate in a search region generated in the previous computation. Because the scale of road network has great influence on path search algorithm, our method which uses road networks inside the search regions for path search outperforms MS method which uses whole road networks for search. The difference between execution times of two methods is clear in every situation of the tests.

(2) The scale of search region in COMS method refers to the size of the region and the number of segments (or nodes) on road network inside the region. The scale of search regions generated in every step in COMS is based on the path length from source point to the candidates. In the situation of target objects distributed uniformly, if there are more targets on the grid, the path length from the source to the nearest target becomes shorter, and the search region also becomes smaller. The path search can be based on small search region, and the search cost is low. For the same number of target objects, a bigger grid leads to a bigger scale of search region, so the execution time on Grid 30×40 is longer than that on Grid 21×28 in the figure.

On the other hand, MS method starts from the overlaying of target objects and the grid, and executes the nearest search based on the whole grid. The NN search time is related to the grid scale, so the speed is faster for a smaller grid. However, the random distributions of more target objects have higher possibility for leading to extreme distributions of target objects. Namely, there are much more target objects locating inside a search region than those with uniform distributions of target objects. The extreme distribution influences the results of the method.

(3) For the same reason as the previous experiments, the search time of MS method is quite affected by the number of target objects and the scale of grid. On the other hand, in COMS method, the search is computed inside search regions, which is only a part of the total grid. Therefore, the difference between the two methods becomes bigger with the increase of the scale of grid and number of target objects.

(4) All of the execution times in our test on real road network are longer than those on grid. This difference comes from the fact that the straight-line distance between objects on the grid is in direct proportion to the path length between the two objects, while on the real map this is not always true. However, as COMS method does not need to overlay all the target objects and the road network, and the search region used in the search process is smaller than the total network, the execution time difference between the two methods becomes bigger on the average with the increase of the scale of road network and the number of target objects. Certainly, when there is an extreme distribution of target objects and an extreme situation of road network, COMS method may be defeated by MS method. For example, the path length from the source to a candidate generated in search process is long enough to generate a search region bigger than the total road map.

5.2 Continuous Nearest Neighbor Search on Road Network

5.2.1 Introduction

Continuous nearest neighbor (CNN) search has been studied in recent years from the computational geometry perspective [86–88]. All their works are based on the straight-line distance between objects. However, in GIS or ITS, the continuous nearest target objects for a predefined route are measured by the path cost from the current position in case that the cost can represent the shortest distance, travel time, etc. For example, the result of "find all my nearest gas stations at any point on my route from place S to place E" is a set of $<point, interval, path>$ triples, such that $interval$ is a sub-route, $point$ is the nearest target object of all points on the sub-route, and $path$ is the shortest path from the sub-route to the $point$.

Furthermore, this kind search may refer to a road network in a wide area where the predefined route crosses. Usually, the whole road network is too large to be stored in the main memory at once. To solve the problem of path search on a large road network, many methods have been proposed [89–92] from viewpoints of partition and precomputation. In these methods, the road network is divided into partitions which are small enough to be resident in the main memory. To accelerate the path search on the partitioned road network, all-pair shortest paths among the boundary nodes and/or among nodes inside every partition are precomputed. These methods are appropriate for the path search between two points on the road network, nevertheless in CNN-search the path search is performed repeatedly and the cost of accessing partitions on disks is expensive. Therefore, this section gives a method to decrease the times of disk access by minimizing the path search region in CNN-search process. In other words, CNN-search may refer to the information about large road network, and the nearest neighbor (NN) search for the intervals on the route may be performed on a relatively small area.

5.2.2 Road Network, Route and Computation Point

A road network with nodes and links representing the cross-points and road segments can be regarded as a graph G: $G = (V, L)$, where V is a set of vertices $\{v_1, v_2, ..., v_n\}$, and L is a collection of lines $\{l_1, l_2, ..., l_m\}$. The predefined route from a start point v_s to an end point v_e is given by an array $Route(v_s, v_e) = \{(v_{r1}, ..., v_{ri}, ..., v_{rm}) | v_{r1} = v_s, v_{rm} = v_e, v_{ri} \in V, ri = r2, ..., rn - 1\}$. A sub-route of $Route(v_s, v_e)$ is defined as $(v_{rl}, ..., v_{rj})$, which overlaps with $Route(v_s, v_e)$. If the target object set is $T = \{t_a, t_b, ...\}$ and $t_i \in T$ with a corresponding node $v_{ti} \in V$, NN for v_{ri} on $Route(v_s, v_e)$ is t_i when the shortest path $Path_{vri\_ti} = \{min(v_{ri}, ..., v_j, ..., v_{ti}) | v_j \in V, ti = t_a, t_b, ...\}$.

Definition 5.2 Node $v_d \in V$ is called the divergence point between $Route(v_s, v_e)$ and $Path_{vri\_ti}$ only if the two following conditions are satisfied:

Fig. 5.4 Road network,
specific route [S, E], and
target objects

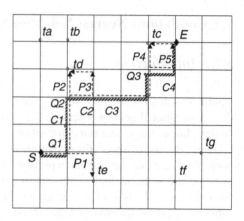

(1) The sub-route $(v_{ri}, ..., v_d)$ of $Path_{Vri\_ti}$ is also a sub-route of $Route(v_s, v_e)$.
(2) The node following v_d along $Path_{Vri\_ti}$ is not on $Route(v_s, v_e)$.

Consider Fig. 5.4: t_e is NN of node S on $Route(S, E)$ and the shortest path from S to t_e is P_1. The divergence point between $Route(S, E)$ and P_1 is Q_1, and is just the point in which the shortest path branches off the route. It is obvious that the points on the sub-route $(v_{ri}, ..., v_d)$ share the same NN t_i. Therefore, there is no necessity to search NN for every point on the sub-route; CNN-search can be regarded as a series of NN-searches for some points on the route; and those points are called computation points.

Definition 5.3 Node $v_c \in V$ is called the computation point of $Route(v_s, v_e)$ only if the two following conditions are satisfied:

(1) v_c is the start point of $Route(v_s, v_e)$.
(2) v_c is on $Route(v_s, v_e)$ and is also a node on the route following a divergence point between $Route(v_s, v_e)$ and $Path_{Vri\_ti}$.

In Fig. 5.4, S is a computation point and C_1 is a computation point following the divergence point Q_1. NN for C_1 is t_d. t_d is also regarded as NN for all the points on the sub-route of $Route(v_s, v_e)$ from the previous divergence point Q_1 (except for Q_1 itself, its NN is t_e) to the following divergence point Q_2. This is because on the real road network we can branch off the route only on cross-point.

5.2.3 Path Search Regions

To solve CNN problem, there are two main issues: one is the selection of computation point on the route; the other is the computation of NN for the computation point. Here, we give two propositions on the road network for nearest object search.

Proposition 5.4 *For a source point S and a target object t, when the length of a path from S to t is r, if any target object is nearer to S than t, it can only be found inside a circle region, denoted as r-region, whose center is S and radius is r (Fig. 5.2).*

This proposition depicts the same fact as Proposition in Sect. 5.1. We leave the proof out in this chapter, as it can be explained by the fact that any road segment outside *r-region* can only lead to a path longer than *r* from *S*.

Proposition 5.5 *For two points S and t on the road network with straight-line distance d, the test of whether there is a path from S to t shorter than r can be based on a path search region, denoted as p-region. The sum of the straight-line distance between any nodes inside this region and S and that between this node and t is not longer than r.*

For an easy description we define coordinates for them in Fig. 5.5:

$$p - region = \{(x, y) | \sqrt{(x + d/2)^2 + y^2} + \sqrt{(x - d/2)^2 + y^2} \le r\}. \qquad (5.6)$$

The origin O is on the center of line $\overline{St}$, the x-axis passes along line $\overline{St}$, and the y-axis is perpendicular to the x-axis on the origin O. Finding the shortest path from S to t is based on the road segments inside the region (as the grey ellipse in Fig. 5.5). This means that if there is any path shorter than r from S to t, then all the road segments on this path could only be found inside *p-region*. Here, we prove that this region is legal, and also the smallest one.

(1) To prove that the region is legal, we give an assumption that there is a point *p1* *(x1, y1)* outside *p-region* locating on a path from *S* to *t* and the length of this path *SP(S, t)* is not longer than *r*:

$$SP(S, t) \le r. \qquad (5.7)$$

Fig. 5.5 Shortest path
search region: *p-region*

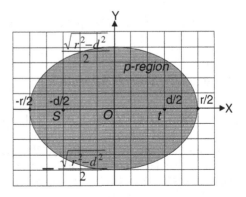

According to triangular inequality, the path length is not shorter than the straight-line distances among S, $p1$ and t:

$$Straight_line_dis(S, p1, t) = \sqrt{(x1 + d/2)^2 + y1^2} + \sqrt{(x1 - d/2)^2 + y1^2}; \quad (5.8)$$

$$Straight_line_dis(S, p1, t) \leq SP(S, t) \leq r; \quad (5.9)$$

and

$$\sqrt{(x1 + d/2)^2 + y1^2} + \sqrt{(x1 - d/2)^2 + y1^2} \leq r. \quad (5.10)$$

By the definition of *p-region*, *p1* is inside *p-region*. This result is contradictory to the assumption.

(2) To prove that the region is the smallest region for the search, it is assumed that there is a point *p1 (x1, y1)* on the boundary of *p-region* and a path from S to t crossing *p1*.

$$Straight_line_dis(S, p1, t) = r; \quad (5.11)$$

By the definition, the length of this path $SP(S, t)$ is not longer than r and $SP(S, t)$ is not shorter than $Straight_line_dis(S, p1, t)$:

$$Straight_line_dis(S, p1, t) \leq SP(S, t) \leq r; \quad (5.12)$$

then

$$SP(S, t) = r. \quad (5.13)$$

This means that any region of smaller than *p-region* may be results in an answer missing and *p-region* is the smallest region for this search.

The region can also be simplified to a rectangle with length r and width $\sqrt{r^2 - d^2}$.

5.2.4 CNN-Search Approach

The problem of CNN-search is to find NN for any point along a specific route on a large road network. NN is the target object with the shortest path from the point on the route. In the ordinary GIS, road network and other map objects are managed in index structures, respectively. For example, the target objects are indexed by R-tree [17]. Therefore, CNN-search is solved by taking the propositions into the following considerations:

(1) *Locating the target objects inside r-region on the road network*
   With the assumption that the road network is too large to be processed in the main memory at once, locating target objects on the road network leads to an expensive process of merging two indexes of road network and target objects. However, based on Proposition 5.4, if a straight-line restriction between the computation point and a NN candidate can be decided, the locating operation can only be performed inside *r-region* for the computation points.
(2) *Searching shortest path from the computation point to NN candidate inside p-region*
   Based on Proposition 5.4, if a path length restriction from the computation point to a NN candidate can be decided, the path search can be performed inside *p-region*.

   The following parts of this section address the methods for deciding *r-region* and *p-region* in CNN-search process.

5.2.4.1 Decision of *r-region* and *p-region*

Observe Fig. 5.6: the predefined route is $Route(S, E)$, and the target object set is T. The first computation point is the start point S of $Route(S, E)$: NN for S is t and the divergence point is q. The second computation point is decided as c, based on Definition 5.3. The path length from c to t is r, where

$$r = Path_{cq} + Path_{qt}. \tag{5.14}$$

It means that t is a NN candidate for c with path length r, the value of $Path_{qt}$ has been computed in the previous NN-search step for S, and the value of $Path_{cq}$ is the curve length between c and q.

   Based on Proposition 5.4, if *r-region* is decided for C with the radius r, the target objects nearer than t can be found only inside this *r-region*. For NN-search of c, only

Fig. 5.6 Search regions generated for a new computation point: NN-search region *r-region* and shortest path search region *p-region*

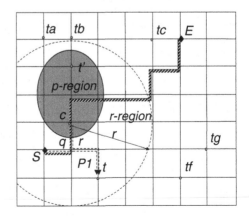

the target objects inside *r-region* are regarded as candidates, and locating candidates on the road network can be performed inside *r-region*.

Based on Proposition 5.5, it can be based on a *p-region* to test whether the path length from c to a candidate t' inside *r-region* is shorter than r: the path length from the current computation point c to NN of the previous computation point is r, and the straight-line distance between c and the candidate node t' is d'. *p-region* for the computation of the shortest path from c to t' is the gray ellipse in Fig. 5.6.

5.2.4.2   Relation Between *p-region* and *r-region*

The selection of a candidate for NN-search is based on the straight-line distance between the computation point and the target objects. Therefore, the next candidate for the computation point is located in a ring, which is between circle d and circle r in Fig. 5.7. After the computation of the shortest path from c to a candidate t', if r' is shorter than r, r' is set as a smaller *r-region* for the following search. And *p-region* for the next candidate becomes smaller, too. With the search steps, as the candidate target is tested one by one in the sequence of d and r is set as the shortest path length up to now, d becomes longer and longer while r becomes shorter and shorter. *p-region* becomes smaller and smaller, too. The area of *p-region* for the candidate t' is simplified as

$$Area(p-region') \propto r \times \sqrt{r^2 - d'^2}. \tag{5.15}$$

p-region is inside *r-region*, and the relation between the areas of them is:

$$\frac{Area(p-region')}{Area(r-region)} \propto \frac{r \times \sqrt{r^2 - d'^2}}{\pi r^2} \leq \frac{\sqrt{r^2 - d'^2}}{\pi r} < \frac{1}{\pi}. \tag{5.16}$$

Fig. 5.7   Relations between *p-region* and *r-region* for one computation point

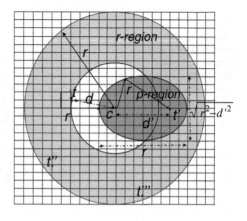

The area of p-$region$ for the next candidate t'' can be simplified as

$$Area(p - region'') \propto r' \times \sqrt{r'^2 - d''^2}. \tag{5.17}$$

As $r' \leq r$ and $d'' \geq d'$,

$$Area(p - region'') \leq Area(p - region'). \tag{5.18}$$

The relation between p-$region$ and r-$region$ is very useful in the CNN computation based on the large hierarchical road network.

5.2.5 Algorithm for Large Hierarchical Road Network

In order to find CNN for a predefined route on the large hierarchical road network, a partition method similar to HEPV [92] is adopted. Roughly speaking, the partitions of road network are regarded as a set of rectangles, e.g., bold-line rectangles in Fig. 5.8. The boundary nodes are pushed to the upper level to form a super graph. All-pair shortest paths among the boundary nodes in the same partition are computed, and the corresponding edges are added to the super graph as *super-links*. c and t' are the current computation point and NN candidate, respectively; and u_i and v_j are boundary nodes in c-$partition$ and t'-partition, respectively. In the super graph, there are *super-links* pre-computed among u_i ($i = 1, 2, ...$) and among v_j ($j = 1, 2, ...$). The gray region in Fig. 5.8 represents current p-$region$. The shortest path from c to t' is:

$$SP(c, t') = min_{u,v}\{SP(c, u) + SP(u, v) + SP(v, t')\}. \tag{5.19}$$

Here, u and v are limited to be inside p-$region$.

p-$region$ may be decreased in the computation steps; thus, the numbers of u and v become smaller and smaller. In this section, we first give a method for selecting the pair (u, v), and then give the algorithm for CNN-search based on a large hierarchical road network.

Fig. 5.8 Situation of
p-region and border nodes
when computation point and
NN candidate belong to
different partitions

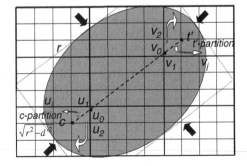

5.2.5.1 Heuristic for Sorting Boundary Nodes

In order to find the shortest path from the source node to the end node, the method proposed in [92] should compare the path lengths for all the pairs of the source and end boundary nodes. In our method, the computation is limited to the boundary nodes inside *p-region*. Furthermore, *p-region* may be decreased during the computation process; and some pairs of (u, v) in the previous *p-region* may not be inside the following *p-region*. *p-region* is decreased along the directions of black arrows in Fig. 5.8. The boundary node on *c-partition*, which is further from u_0, has a higher possibility to be ignored in the following search. So, we select boundary nodes in the sequence of the distance between them and a special node (the cross-node of the corresponding boundary and line $\overline{ct'}$). In Fig. 5.8, u_i (v_j) is sorted on the distances between u_i (v_j) and u_0 (v_0). The directions of white arrows in Fig. 5.8 depict this situation.

5.2.5.2 Algorithm for CNN-Search on Large Hierarchical Road Network

Here, the algorithms of CNN-search and NN-search with the shortest path search are described based on *p-region*. As CNN-search is divided into a series of NN-searches for computation points on the route, the algorithm of CNN-search takes the responsibility of generating the computation point and initializing NN-search region. At the beginning, the computation point is the start point of the route, and the search region for NN-search is set to the Maximum value. The Maximum value could be specified by the user (e.g., less than 2 km away from the route) or determined by some propositions. The following steps run repeatedly to compute the result of triples for every computation point by calling NN-search procedure, and generate the computation point and corresponding search region *r-region*. The algorithm is given as follows:

Algorithm 2: CNN-search

---
    input : route [S, E], target object set T
    output: Result set of triples $\{ < point, interval, path >, ... \}$
1 **begin**
2 | Initialize: set first computation point: $CP = S$;
3 | Set NN search region for $CP : r = Max$;
4 | **while** $CP \neq E$ **do**
5 | | Call NN-search with CP, r;
6 | | Get a triple of $< t, [CP, q], Path_{CPt} >$;
7 | | Replace interval $[CP, q]$ with $[q_{pre}, q]$, insert $< t, [q_{pre}, q], Path_{CPt} >$ into Result set;
8 | | Generate next computation point CP: CP = next intersection from q along route;
9 | | Set NN-search region for CP: $r = Path_{CPq} + Path_{qt}$;
10 | **end**
11 **end**

NN-search process is based on R-tree index and a priority queue *Queue*. *Queue* is used to record the intermediate results: the candidate targets or the internal nodes of R-tree. The key used to order the elements on *Queue* is the straight-line distance of R-tree node and the path length computed for target object. *Queue* is initialized as a node of R-tree, which overlaps with the search region. When a target object turns on the head of the priority queue, it becomes the candidate for further computation: the path length is computed for the candidate based on the search region. If the path length is smaller than the key of the head element of the queue, the candidate is the result. Otherwise, when the path length is smaller than the radius of search region, the search region is reset with the new length as the radius. The value of radius is decreased by keeping step with the ongoing path length computation for the candidates, and the search region is adjusted until there is no candidate inside the search region.

NN-search algorithm is given as Algorithm 3.

Algorithm 3: NN-Search

    input : route [S, E], target object set T; source point CP and search region r
    output: a triple $< point, interval, path >$

1 **begin**
2   Initialize: priority queue *Queue*;
3   Locate first node overlapping with region r on R-tree, compute straight-line distance d between it and CP, and insert node into *Queue*;
4   **while** *Queue* $\neq$ *NULL* **do**
5   **if** *head of Queue is leaf node of R-tree* **then**
6   Initialize *p-region* with *r* and *d*;
7   Call SP-search (Algorithm 4) to compute shortest path SP from CP to it;
8   Insert result to *Queue*;
9   **if** *SP is smaller than r* **then**
10   Reset *r* with *SP*;
11   **end**
12   **else**
13   Compute straight-line distance between them;
14   Insert result to *Queue*;
15   **end**
16   **if** *head of Queue is leaf node with computed shortest path* **then**
17   Find divergence *q* of $Path_{CP_t}$ and route;
18   Remove head of *Queue*;
19   Return result of triple $< t, [CP, q], Path_{CP_t} >$;
20   **end**
21   Remove head of *Queue*;
22   **end**
23 **end**

Algorithm 4: SP-search

---

    input : source point CP, candidate point t and p-region
    output: shortest path length
1 **begin**
2 | Locate CP and t to partitions on ground level;
3 | Compute out u_0 and v_0 of corresponding partitions;
4 | Do steps 5 to 8, until there is no new pair of (u_i, v_j);
5 | Sort u_i (v_j) in order of straight-line distance between u_i (v_j) and u_0 (v_0);
6 | Select pair of (u_i, v_j);
7 | Compute SP with path length of (CP,u_i), (u_i,v_j) and (v_j,t'), If SP is less than r then reset
       | $p - region$;
8 | Return r as shortest path length;
9 **end**

5.2.6 Evaluation

The experiment is developed in Java on an SGI O2 R5000 SC 180 entry-level desktop workstation. The system manages road maps of a part of Aichi Prefecture, Japan. The map is divided into map pages with the same size of 2,000 m × 1,500 m. There are 42,062 nodes and 60,349 road segments on the road network. The road network is represented by spatial objects of nodes and links. By using MP-method, 2,680 boundary nodes are generated.

5.2.6.1 Characteristics of Partitioned Road Network

In this subsection, the features of maps partitioned by MP-method are compared with those by Hierarchical Encoded Path View (HEPV) method [92]. These features lay the foundation for understanding the behaviors of two methods with respect to retrieval, precomputation and storage.

For a n $(=m \times m)$ grid graph, the graph is partitioned into p map pages. Thus, there are n/p nodes including up to $4\sqrt{n/p}$ boundary nodes in every map page. Super-links generated for a map page are up to $16n/p$. The total boundary nodes are up to $2\sqrt{np}$ ($\approx 2\sqrt{n}(\sqrt{p}+1)$), and super-links are up to $16n$. While, in HEPV method, a graph is first divided into partitions, and then boundary nodes are pushed to the second level to form a super graph. The all-pair shortest paths within every partition (called FPV), and all-pair shortest paths among boundary nodes (called HPV) at all levels are computed. The size of a partition may be smaller than that of our map page. Based on the test results in [92], the best performance is emerged when there are about 200–300 nodes in every partition and at least 3 levels for a road network with 10,000 nodes. Considering the real dataset with more than 40,000 nodes used in the test, the average number of nodes in a map page is about 800: thus we roughly divide a map into 4 partitions, and create 4-level HPV. That is to say, the previous grid is divided into $4p$ partitions in HEPV method. Then, each partition has $n/4p$ nodes

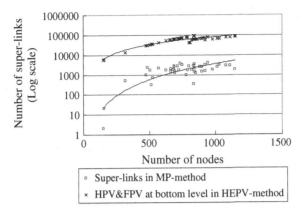

Fig. 5.9 Relation between number of nodes and that of super-links in map pages

including up to $2\sqrt{n/p}$ boundary nodes. The number of total boundary nodes at the bottom level is up to $4\sqrt{np}$ ($\approx 2\sqrt{n}(2\sqrt{p}-1)$). The total FPV at the bottom level is up to $n^2/4p$ ($\approx n(n/4p-1)$). The total HPV computed for the bottom level is up to $4n - 2\sqrt{np}$ ($= 2p\sqrt{n/p}(2\sqrt{n/p}-1)$). The total pre-computed records (including FPV and HPV) for the bottom level are up to $n^2/4p + 2n$. When a road network is large, the number of super-links ($16n$) generated in MP-method is quite smaller than those of FPV and HPV in HEPV method.

The comparison between the number of super-links of every map page in MP-method and the number of (FPV + HPV) for every bottom partition in HEPV method is given in Fig. 5.9. In this test, the map page with less than 200 nodes is regarded as a partition in HEPV method, and other map pages are regarded as four bottom partitions in HEPV method. Figure 5.9 shows that the precomputed links of HEPV method is about 100 times of those in our method. Just because the super graph of HEPV is huge, the cost for path search crossing several partitions is high.

5.2.6.2 Characteristics of Search Regions

For CNN-search, two kinds of search regions are used: one is *r-region* for searching candidate targets; and another is *p-region* for the shortest path search from the source point to the candidate target. While, in HEPV method the shortest path search is done with all levels and all pairs of boundary nodes. Here, the size of *p-region* is analyzed and the test results are given on the comparison of disk accesses.

For the previous grid, suppose that the grid is in two-dimensional interval $[0, m] \times [0, m]$, then there are N_t ($=m_t \times m_t$) targets distributed uniformly on the grid. The number of targets inside *r-region* (with radius r) is up to $\pi r^2 N_t/N$. The straight-line distance (d) from the source point to the first candidate target tested inside *r-region* is up to $\sqrt{n/2N_t}$. In the large road network of real world, we can observe that the path length between two nodes is usually not longer than two times of the straight-line

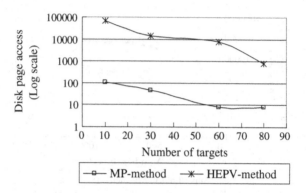

Fig. 5.10 Comparison of disk page access in MP-method and HEPV-method

distance between them. That is to say, $r < 2d$. The area of *p-region* is up to

$$r \times \sqrt{r^2 - d^2} = \sqrt{3}n/N_t. \qquad (5.20)$$

Thus bottom nodes inside *p-region* is up to $\sqrt{3}n/N_t$. Considering the angle between the rectangle of map page and *p-region*, the maximum number of referred map page is smaller than $2\sqrt{2}p/N_t$, and the boundary nodes inside the region is less than $2\sqrt{n/N_t}$.

As the shortest path search is based on a considerable small number of nodes, the disk access in the search process is greatly decreased. Though the shortest search inside a map page or between boundary nodes of different map pages should be computed by on-demand, such a search is limited to *p-region* and the efficiency is assured. The comparison of disk access in CNN-search by using two methods is given in Fig. 5.10. The targets used in the test are point objects with uniform distributions on the road network. The number of targets is varied from 10 to 80. The workstation has 64 MB of main memory and 80 GB of disk. The page size is 4 KB for disk I/O. In HEPV method, the size of each record (from-node, to-node, partition-id, next-hop, weight) for HPV and FPV is 40 Bytes ($8 * 5$) , which leads to 100 records per disk page. In our method, the record for road segment is (map-page-id, from-node, to-node, weight), the size of a record is 32 Bytes ($8 * 4$) and leads to 125 records per disk page; the tuple for super-link is (from-node, to-node, map-page-id, next-hop, weight), whose size is 40 Bytes ($8 * 5$) and leads to 100 records per page. In Fig. 5.10, when there are fewer targets on the road network, the difference between disk access of MP-method and that of HEPV method is larger. This is because the average path length from the the predefined route to targets becomes longer and the HEPV method requires expensive distance computations by incurring extensive buffer thrashing.

There is a comparison with different methods for CNN-search in Fig. 5.11 and Fig. 5.12. The x-axis is set as the density (= the number of targets/the number of nodes on road network) of the target objects on the road network; and the y-axis is set as the average NN-search time of every computation point during CNN-search. In

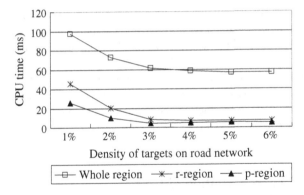

Fig. 5.11 CPU time of finding NN for every computation point: all the road network can be stored inside the main memory at once

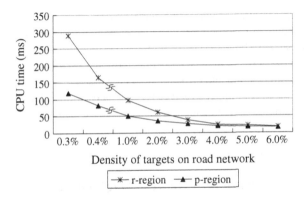

Fig. 5.12 CPU time of finding NN for every computation point: all the road network cannot be stored inside the main memory at once

Fig. 5.11, the road network is set as a grid (30 × 30), the target objects are distributed on the grid uniformly with the density from 1 to 6%. Because the whole network can be stored into the main memory, the comparison is made among three methods: (1) Dijkstra's algorithm for path search based on the whole network; (2) path search based on *r-region* (the method proposed in Chap. 6); and (3) path search based on *p-region* (proposed in this chapter). It is easy to understand the result, because the size of the search regions is different: *p-region* is smaller than *r-region* and *r-region* is smaller than the whole region (the whole network), absolutely; method (3) is more efficient than method (2); and method (2) is more efficient than method (1). When there are more targets on the road network (the density is high), the average path length from the route to the targets is short, the difference between the sizes of *r-region* and *p-region* is small, and the gap between the search times is small, too. However, the methods using search regions are quite more efficient than that based on the whole region in any situation. The similar results can be seen in Fig. 5.12. When the road network cannot be stored inside the main memory at once, the path search cannot be based on the whole network. The comparison is made between the searches using *r-region* and *p-region*. The size in *p-region* is quite smaller than that in *r-region*, and the use of *p-region* shows better property in any situations, especially, when the density of targets on the road network becomes quite low (less than 1.0 %) the performance is good. In practice, the number of target objects is quite less than the nodes of the road network in common sense.

5.3 Reverse Search Method of CNN

5.3.1 Introduction

ITS applications are often based on the current travel time, congestion, restrictions and other attributes of road network. The *super-node* method decreases the redundancies in the database by adopting a complex node representation. Though the total information for the network may be more than those in [53] method, it is easy to integrate the traffic information and the static road network. This method does not injure the stability of the spatial index structure for road network. In this section, we focus on the search method on the transportation network. We also use the CNN-search as an example.

5.3.2 Temporal Continuous Nearest Neighbor Search

In this section, a method for CNN-search along a predefined route is proposed based on a dataset, denoted as *super-node* dataset, which is generated by the *super-node* representation method. The predefined route from a start point v_1 to an end point v_n is given by an array $Route(v_1, v_n) = (v_1, v_2, ..., v_{n-1}, v_n)$, and the target object set $\{t_a, t_b, ...\}$ is managed by a spatial index structure (e.g., R-tree [17]). We center on the *super-node* repressentation method and its influence on CNN-search. The *super-node* dataset consists of information about road network and traffic cost on the network.

We make observations of the *super-node* dataset in CNN-search process:

(1) Every vertex in the *super-node* dataset keeps the cost information of the possible out-arcs, so the cost of traveling from a vertex v_i on $Route(v_1, v_n)$ to the following vertex v_{i+1} is kept on vertex v_i and denoted as $v_i.cost_{i+1}$. If the nearest neighbor (NN) of v_{i+1} is known as t_{i+1} with $cost(v_{i+1}, t_{i+1})$, the cost of traveling from v_i to its NN t_i is no larger than a value $Cost\text{-}limit(v_i)$, which is computed by:

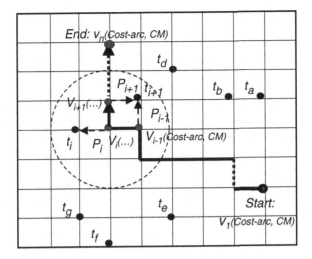

Fig. 5.13 Predefined route and NN for v_i

$$Cost\text{-}limit\ (v_i) = v_i.cost_{i+1} + cost(v_{i+1}, t_{i+1})$$

Cost-limit (v_i) is used to set a region for the NN-search of v_i (e.g., in Fig. 5.13), NN of v_i can only be found inside the dotted circle region. The region is defined as a circle with the radius of *Cost-limit* (v_i) and center of v_i.

(2) The nearest target object t_{i+1} of v_{i+1} is also the nearest one on the possible paths from v_i via v_{i+1}. In other words, t_{i+1} is the nearest one found on a path from v_i via $v_i.out_{i+1}$. If there is any object being nearer to v_i than t_{i+1}, the shortest path from v_i to this object does not pass through v_{i+1}. Certainly, it is possible that there is a path from v_i to t_{i+1} via v_j ($j \neq i + 1$), which is shorter than *Cost-limit* (v_i). This situation is depicted in Fig. 5.13, where v_{i-1} and v_{i+1} share the same NN t_{i+1}, but there is no overlap between the two paths p_{i+1} and p_{i-1}.

Based on the previous observations, it can be concluded that:

(1) The path length from v_i to NN t_{i+1} of v_{i+1} can be set as a limit for NN-search of v_i.
(2) NN-search of v_i can be executed along the out-arcs of v_i except for $v_i.out_{i+1}$.

Here, a simple proof is given for these conclusions:

(1) It is proven that t_{i+1} is a candidate of NN-search of v_i. Based on the definition of $Route(v_1, v_n)$, v_i and v_{i+1} are along the same route, so there is an out-arc of v_i leading to v_{i+1}. Being NN object of v_{i+1}, t_{i+1} can also be reached from v_i via v_{i+1}. Therefore, t_{i+1} maybe NN of v_i, too.
(2) It is proven that *Cost-limit* (v_i) is the shortest one from v_i to any object via v_{i+1}. If there is another object t' with a path shorter than that of v_{i+1} via v_{i+1}, then t' is also nearer to v_{i+1} than t_{i+1}. This contradicts to the promise that t_{i+1} is NN of v_{i+1}.

5.3.3 Algorithm Description

A method for CNN-search along $Route(v_1, v_n)$ is proposed. This method begins from the end vertex of this route, and searches NN for every vertex in the reverse order of this route.

Firstly, the method searches t_n for the end vertex v_n, and then generates a search limit for the next computation vertex v_{n-1} based on the previous result, and checks whether there is an object nearer to v_{n-1} via the out-arcs of v_{n-1} except for $v_{n-1}.out_n$. These steps run in cycle until the computation vertex is v_1. This method is correct, based on the previous observations.

NN-search for every vertex can be realized by adopting a priority queue to maintain the current frontier of the search. Any vertex with a higher cost from v_i than the limit value is not inserted into the queue. By expanding the vertex on the head of the queue, the algorithm ends when the head vertex connects to a target object.

An example for NN-search of v_i is given in Fig. 5.14. NN of v_i is to be searched inside the dotted region. There are assumptions that every grid represents a unit of cost; right-turn and U-turn are forbidden on v_i; and no restrictions are imposed on vertex $v_{4,3}$. The search for v_i begins from the following possible out-arcs of v_i: here, $v_i.out_{4,3}$. As there is no target object connecting to $v_{4,3}$, and the cost from v_i to $v_{4,3}$ is no larger than $Cost\text{-}limit$ (v_i), the search expands the vertex $v_{4,3}$ using the width-first method. The vertex $v_{4,2}$ connecting to a target object t_i with the lowest cost is found, and the object t_i is regarded as NN of v_i. The main steps in this algorithm are given as Algorithm 5.

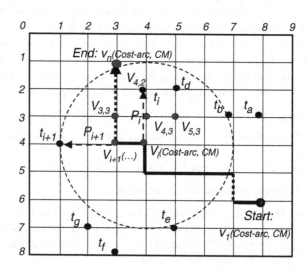

Fig. 5.14 NN-search for v_i with a limit

Algorithm 5: NN-search

---

input : $Route(v_1, v_n)$, target object set T, source point v_i, and
  $< t_{i+1}, interval_{i+1}, path_{i+1}, cost(v_{i+1}, t_{i+1}) >$
output: a triple $< target, interval, path, cost >$

1 **begin**
2 | Initialize priority queue $Queue$;
3 | $Cost\text{-}limit (v_i) = v_i.cost_{i+1} + cost(v_{i+1}, t_{i+1})$;
4 | Sort v_i with $(cost = 0)$ and t_{i+1} with $(cost = Cost\text{-}limit (v_i))$ into $Queue$;
5 | **while** $Queue \neq NULL$ **do**
6 | | **if** *the head of Queue is v_k not connecting to a target object* **then**
7 | | | Expand possible *out-arcs* of v_k: $v_k.out_p$, where $C_{kp} \neq 1$ in $CM(v_k)$;
8 | | | Compute the cost from v_i up to v_p;
9 | | | **if** *the cost is smaller than Cost-limit (v_i)* **then**
10 | | | | Sort v_p into $Queue$;
11 | | | **end**
12 | | **end**
13 | | **if** *the head of Queue is a node connecting to a target object t_i* **then**
14 | | | Return result of triple $< t_i, interval_i, Path_i, cost(v_i, t_i) >$
15 | | **end**
16 | | Remove the head of $Queue$;
17 | **end**
18 **end**

5.3.4 Evaluation

Evaluations are made based on two kinds of representation methods and with different algorithms for CNN-search. The comparison is made on the average CPU time for searching a nearest neighbor for a computation point on the predefined route. By varying the distributions of target objects on the road networks, the search costs in different situations are compared.

The basic road map is indexed by MOR-tree, and the information about nodes is managed in disk pages. The targets used in the test are point objects with uniform distributions on the road network, which is managed by another spatial index. The number of targets is varied from 3 to 12 % of the nodes on the road network. Our workstation has 64 MB of main memory and 80 GB of disk. The page size is 4 KB for disk I/O. The record size of *super-node* (i.e., a cross node on road network, with four $(v_i, cost)$ and a 16 *Constraint-matrix* elements) is 72 Bytes $(4 * (6+8) + 1 * 16)$, which leads to 56 records per disk page. In node-link method, the record for road segment is (from-node, to-node, weight), which is 24 Bytes $(8 * 3)$ and leads to 170 records per disk page. The total number of nodes N_{num} is 42,062 and the number of links L_{num} is 60,349 in the basic road map. The average traffic arcs connecting to a node is about 2.87 $(=2 L_{num}/N_{num})$.

The test results are given in Fig. 5.15. In the figure, x-axis represents the density of targets on the road network, which is the ratio of the targets' number (T_{num}) to the nodes' number (N_{num}) in basic road map; y-axis represents the CPU time of

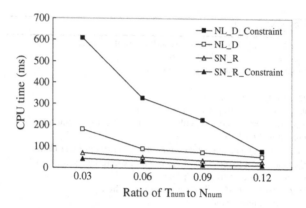

Fig. 5.15 Average CPU time of finding one target in CNN-search

NN-search for every computation-point on the predefined route. D represents that CNN-search is performed by adopting Dijkstra's algorithm for NN-searches. When the traffic cost is set as the length of road segment, NN-search is based on a proper search region, which is a circle; and R represents the reverse search method proposed in this chapter. D algorithm is executed based on the datasets generated by NL method with/without constraint; R algorithm is done on the dataset generated by SN method with/without constraint.

In Fig. 5.15, we can observe that our *SN_R* methods outperform those of *NL_D* at any situation, especially, when there are traffic constraints on the networks. This is because:

(1) In this method, NN-search only expands the nodes on road map in possible directions when there is any traffic constraint on the node. With our reverse CNN-search method, one more direction of the possible directions is decreased, and the following nodes on that direction does not need to be tested. The search cost is shrunken.

(2) In *NL_D* method, Dijkstra's algorithm is executed inside a proper search region. Therefore, NN search process includes the steps of searching regions generation, distance matrices creation for networks inside the search regions and the path searches by using Dijkstra's algorithm based on the matrices. When there are more targets on the network, the search region becomes small, and the scale of the network inside the search region becomes small, too. So, the search is faster when there is a higher ratio of T_{num} to N_{num}. The cost of matrix creation and path search is related to the number of nodes and arcs inside the search region. When there is any traffic constraint, the number of nodes and arcs increases, and also the cost increases, accordingly. In Fig. 5.15 the CPU time of *NL_D* is faster than that of *NL_D_Constraint*, and both of them decrease when there are more targets on the network. The search region can be generated only when there are direct relations between the travel cost and the length of road segment. So the comparison is only done on this assumption. The analysis of other situations is discussed in the next subsection.

5.3.4.1 Analysis of Traffic Cost

The abstract cost is used in the previous sections. In this subsection, an analysis is given from a viewpoint of providing concrete examples of *cost*: when there is a uniform speed of traffic on the road network, *cost* is the length of the road segment; otherwise, *cost* is the travel time for every traffic arc on the road network. Certainly, there are other kinds of *cost*: for example, the toll of a path. This method supports the search based on all these *cost* definitions.

The discussion and examples used in the previous sections can be regarded as the traffic arcs with the assumption that *cost* is equal to the length of the road segment, implicitly. If *cost* is the travel time on the traffic arc though the region may be not a circle on the road map, it is sure that a region can be generated with the same method and also NN-search for every vertex can be executed using the same program. This is because the region is actually realized by adopting the priority queue ordering on *cost*. The values of the turn cost can be used naturally in the process of search algorithm.

On the other hand, either kind of cost is adopted in the dataset, and the quantity of information on every vertex keeps the same. Therefore, when the road map is managed by some spatial index (e.g., R-tree, MOR-tree for multi-scale road network), *Cost-arc* and *Constraint-matrix* associated to a vertex are stored into a fixed space of specific disk page. The update for traffic information does not injure the stability of the spatial index.

5.4 Forecasting Aggregate Query on Road Network

5.4.1 Introduction

Forecasting Aggregate Query is defined as follows: Given the forecasting time period $[t_0, t_1]$ and the query area $q = ([x_0, x_1], [y_0, y_1])$ (where t_0 is greater than the current time, $[x_0, x_1], [y_0, y_1]$ are a two-dimensional spatial coordinates description of the query area), it forecasts the approximate total number of moving objects on all sections within the query region q in the period of time. Mainly discusses two cases: (1) If $t_0 == t_1$, it expresses forecasting aggregate query about moving objects on the query section of road networks in a single moment; (2) If $t_0 < t_1$, it expresses the sum of forecasting results obtained about forecasting aggregate query of moving objects on the query section of road networks within the plurality of discrete moment in the time period $[t_0, t_1]$.

At present, domestic and foreign dynamic traffic flow forecasting theory researches are still in development stage. They have not formed a more mature theoretical system, especially for short-term traffic flow forecasting methods research. It is impossible to obtain satisfactory results. The traffic flow short-term forecasting forecasts the traffic flow during a few hours or even minutes later. It will be affected by many factors and cannot produce good forecasting effects, such as random events interference. The uncertainty about short-term traffic flow is stronger than long-term traffic flow, and the formers regularity is more obvious than the latters. There are many ways about short-term traffic flow forecasting. The main models are: time series models [93], multivariate linear regression models [94], Kalman filtering model [95]. Multiple linear regression model studies interdependence between a number of explanatory variables and the dependent variable. It estimates the value of another variable by using one or more variables. Inspite of the convenient calculation, it creates the model with a relatively simple consideration. The forecasting accuracy is low, and with the forecasting interval shortened, the forecasting accuracy will be reduced. It is not suitable for short-term traffic flow forecasting. Kalman filter model was proposed in 1960s, which has been successfully used in the field of transport demand forecasting. The model is based on the model proposed in the filtering theory by Kalman. It is a matrix iterative parameter estimation method for linear regression model, whose selection is flexibility and precision is higher. But this method needs to do a lot of operations about matrix and vector, resulting algorithm is relatively complex as well. It is difficult to apply it to real-time online forecasting fields. Time series model has been widely used in weather forecasting, hydrological forecasting, stock market prediction, data mining and other fields. The theory is relatively mature, and it is a more promising short-term traffic flow forecasting method. There are Exponential Smoothing (ES) forecasting method based on the DynSketch [83] index structure and SAES forecasting method based on the DS index structure [96, 97], which are aggregate forecasting techniques based on road network data stream aggregate index structures. ES forecasting method is simple and requires only two values, but this method cannot adjust smoothness exponential value adaptively according to the change of data distribution. Self-Adaptive Exponential Smoothing (SAES) forecasting model is the cubic exponential smoothing method of ES. The calculation is larger and more complex. The Modified Histogram (MH) is an improved histogram method by dividing the query area to avoid some unnecessary drawbacks, and to improve the quality of the approximation. The Smooth Transition Exponential Smoothing (STES) method obtains approximate aggregate queries results by less storage space and time consuming.

5.4.2 Exponential Smoothing

Exponential smoothing (ES) is a well-known forecasting method in time series analysis. For example, we set the current time to 0, and process the query $q(q_R, 1)$ to predict the number of objects in q_R at the next timestamp. According to ES method, the estimated result S_1 of the query q can be modeled as a time series:

$$S_1 = aS_0 + a(1-a)S_{-1} + a(1-a)^2 S_{-2} + \cdots + a(1-a)^n S_{-n} \qquad (5.21)$$

where S_0 is the actual result of the PT query $(q_R, 0)$ (at the current time), S_{-i} is the actual result of the HT query $(q_R, -i)$ at the i-th previous timestamp, n is the length of previous history that will be used for prediction, and a is the smoothing parameter in the range $(0, 1)$. The equation is based on the idea that recent timestamps are more important for prediction than older ones. The relative weight is adjusted by the value of a: as it approaches 1, the significance of the most recent timestamps increases.

5.4.2.1 Predictive Aggregate Queries Based on DynSketch

DynSketch index is an efficient spatio-temporal structure for aggregate queries over moving objects. Based on this structure, we use Exponential Smoothing (ES) method to predict the number of moving objects in road networks. The steps of predictive aggregate queries over moving objects are in Fig. 5.16.

Based on the DynSketch index structure, first, we use *agg_qry()* function to estimate the past and present number of the moving objects in road networks. Then, we use ES method to predict the future number of moving objects, that is *Pre_Agg*(see Algorithm 6). In ES method, the past and current data is the actual result. As to obtain actual results needs extreme space and time consumption, different from ES, in our method, we use the approximate result of the least recent timestamp to predict.

In algorithm *Pre_Agg()*, we can do predictive aggregation over moving objects at one timestamp or during a query interval. No matter which situation is, firstly, we need to estimate the number of moving objects at present (line 2) and during the history timestamps (line 4–line 7) using DynSketch index. Then we need to use Eq. (5.21) (line 6) to predict the result at the time after present time. If we want to predict the result at a future time which is several timestamps after the present time, we have to use step by step technique, that is

$$S_{pre+1} = aS_{current} + (1 - a)S_{pre} \qquad (5.22)$$

(line 12, line 18). If we want to predict the result at one timestamp (line 10), we only need to get the result of predictive aggregation at one predictive timestamp (line 11 and line 12), else we need to get the sum of all results at all the timestamps during the query interval (lines 15–21). For example, there is a query: the query interval is [3, 4], the query region is in Fig. 5.17. Based on the DynSketch index structure,

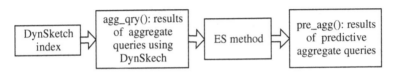

Fig. 5.16 Steps of predictive aggregate queries

Fig. 5.17 Query Region

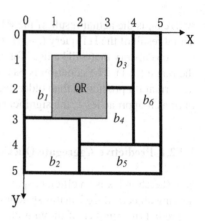

firstly, we need to do *agg_qry()* function to get the estimate result of moving objects in this query region at timestamp 1 and timestamp 2. Then, we will use Eq. (5.21) to get the predictive aggregate results at timestamps 3 and 4 respectively. Finally, we get the sum of the results at timestamps 3 and 4.

Algorithm 6: Pre_Agg($root$, x_0,x_1,y_0,y_1,t_0,t_1)

    input : *root*: the root of DynSketch index, (x_0,x_1,y_0,y_1): the query region, (t_0, t_1): the predictive aggregate timestamp;

1 **begin**
2   s_0 = *agg_query*($root$,x_0,x_1,y_0,y_1,t_0-1,t_0-1);   /* present aggregation query */
3    $s = 0$;
4   **for** $i = 0;i<hislength;i++$ **do**
5       $s[t_0 - 2 - i]$ = *agg_query*($root$,x_0,x_1,y_0,y_1,t_0-2-i,t_0-2-i);
6       $s = s + a * (1 - a)^{i+1}$*$s[t_0$-2-i];
7   **end**
8    $s[t_0]$ = a*s_0 + s;
9   pre_agg = a*s_0 + s;   /* result of prediction at one timestamp */
10   **if** $t_0 == t_1$ **then**
11   **for** $j = timestamp; j > 0; j - -$ **do**
12         $s[t_0 + timestamp - j + 1]$ = a*s_0+(1-a)*$s[t_0$+timestamp-j];
13   pre_agg = $s[t_0$+timestamp];
14   **end**
15   **else**
16   timestamp = $t_1 - t_0$;
17   **for** $j = timestamp; j > 0; j - -$ **do**
18         $s[t_0 + timestamp - j + 1]$ = a*s_0+(1-a)*$s[t_0$+timestamp-j];
19   pre_agg = pre_agg+$s[t_0 + timestamp - j + 1]$;
20   **end**
21   **end**
22 **end**

5.4.3 Self-Adaptive Exponential Smoothing

Choice of exponential smoothing method should adapt to properties of different time series. If there is no significant change in the time series, linear exponential smoothing model (which is also called as exponential smoothing or ES) will be chosen; the time series' change shows linear trend, quadric exponential smoothing model will be chosen; if the time series of change shows parabolic trend, cubic exponential smoothing model [98] will be chosen. According to the analysis in data distribution of road networks will have some volatility, and data streams of peak period in urban road networks are a good verity. Therefore, we may take cubic exponential smoothing model into consideration to optimize prediction calculation on the data streams of road networks.

According to the analysis of traditional exponential smoothing model, we put forward the concept of self-adaptive cubic exponential smoothing which can adapt smoothing weight to time series, and we also call this model as self-adaptive exponential smoothing (SAES). At first, we take a research on cubic exponential smoothing model, whose prediction model is shown in the Eq. (5.23):

$$Y_{t+T} = a_t + b_t T + c_t T^2 \tag{5.23}$$

Y_{t+T} is the target of prediction (prediction value at time $t+T$), t is the time series, T is the predictive time range, and a_t, b_t, c_t are expressed as linear, quadric, cubic prediction parameters. According to the Eq. (5.23), the equations for traditional cubic exponential smoothing are shown in Eq. (5.24):

$$S_t^1 = \alpha X_t + (1 - \alpha) S_{t-1}^1$$
$$S_t^2 = \alpha S_t^1 + (1 - \alpha) S_{t-1}^2$$
$$S_t^3 = \alpha S_t^2 + (1 - \alpha) S_{t-1}^3 \tag{5.24}$$

In the Eq. (5.24), S_t^1, S_t^2 and S_t^3 are expressed as linear, quadric, cubic exponential smoothing value, α is the static smoothing parameter, and X_t is the actual value at time t. And then the prediction parameter is shown in Eq. (5.25).

$$a_t = 3S_t^1 - S_t^2 + 3S_t^3$$
$$b_t = \frac{\alpha}{2(1 - \alpha)^2} [(6 - 5\alpha)S_t^1 - 2(5 - 4\alpha)S_t^2 + (4 - 3\alpha)S_t^3]$$
$$c_t = \frac{\alpha^2}{2(1 - \alpha)^2} [S_t^1 - 2S_t^2 + S_t^3] \tag{5.25}$$

In the cubic exponential smoothing model, the smoothing parameter α is static and is difficult to adapt to the changes of time series, which is the same problem as ES model. The initial value of smoothing parameter is hard to determine [99]. In Eqs. (5.24) and (5.25), we can see that α's value is always a constant value in the calculated process. In terms of the original sequence of ups and downs, even if it finds a suitable value of α for the previous sequence, it will be not necessarily suitable for smoothing and prediction at later periods. For most of time series, randomicity makes many problems in which there is no actual constant value to match the applications all the time. Especially for the road networks' data streams, uncertainty is more obvious. Thus, there will be a clear prediction error and even a serious distortion if the traditional exponential smoothing model is used for prediction. Therefore, we consider giving up a fixed value of α and construct a value $\alpha(t)$ which can adjust itself with changes over time. At first, we change α to $\alpha(t)$ in Eq. (5.24) and get the following equations:

$$S_t^1 = \sum_{i=0 \to t} \alpha(t)(1 - \alpha(t))^{t-i} X_t + (1 - \alpha(t))^t S_0^1$$

$$S_t^2 = \sum_{i=0 \to t} \alpha(t)(1 - \alpha(t))^{t-i} S_t^1 + (1 - \alpha(t))^t S_0^2$$

$$S_t^3 = \sum_{i=0 \to t} \alpha(t)(1 - \alpha(t))^{t-i} S_t^2 + (1 - \alpha(t))^t S_0^3 \qquad (5.26)$$

Because the forms of three equations in Eq. (5.26) are basically the same, here we set $\Psi = \frac{\alpha(t)}{1-(1-\alpha(t))^t}$. Ψ_t is the function of time t. When $0 < \alpha < 1, t > 1$, we can know $0 < \Psi_t < 1$; when $t = 1$, we may get $\lim_{t \to 1} \Psi_t = 1$, and we can make $\Psi_t = 1$ which makes Ψ_t satisfy the condition of smoothing parameter. So we get a new equations:

$$S_t^1 = \Psi_t X_t + (1 - \Psi_t) S_{t-1}^1$$

$$S_t^2 = \Psi_t S_t^1 + (1 - \Psi_t) S_{t-1}^2$$

$$S_t^3 = \Psi_t S_t^2 + (1 - \Psi_t) S_{t-1}^3 \qquad (5.27)$$

The corresponding prediction coefficients are changed as follows:

$$a_t = 3S_t^1 - S_t^2 + 3S_t^3$$

$$b_t = \frac{\Psi_t}{2(1 - \Psi_t)^2}[(6 - 5\Psi_t)S_t^1 - 2(5 - 4\Psi_t)S_t^2 + (4 - 3\Psi_t)S_t^3]$$

$$c_t = \frac{\Psi_t^2}{2(1 - \Psi_t)^2}[S_t^1 - 2S_t^2 + S_t^3] \qquad (5.28)$$

Thus the new adaptive exponential smoothing prediction model (SAES) is constituted by Eqs. (5.23), (5.27), and (5.28). As the new model does not need to estimate the initial values of x_0 and s_0^1, it can smooth X_t and S_t^1 directly. So it deals with the difficulty of determining the initial value and avoids the disadvantage of selecting the initial value of smoothing manually.

5.4.3.1 Architecture

The ultimate goal of this section is to establish an appropriate prediction model to adapt to the practical application of the user's query needs. For example, "query the number of vehicles in the road segment within future ten minutes". This section does not make a further study on aggregate query technique of spatio-temporal data streams, and just uses the aggregate query method of DynSketch; it also can interpret the differences of performance between ES model and SAES model in the experiment section. Here we take the basis of the aggregate results to establish an appropriate prediction system model which is shown in Fig. 5.18.

Since traffic flow data are stored in "Aggregate Index Architecture" with the form of aggregation, the user makes a query according to his (her) need, and SAES module gets the appropriate results from "Aggregate Index Architecture" based on this query conditions, then the algorithm of SAES itself obtains the prediction results, and the final results are sent to the user.

"Prediction query" is defined as the query whose time bucket is $t = [T_1, T_n]$ (T_0 is current time, and $T_1 > T_0$), prediction query region is $q = ([X_0, X_1], [Y_0, Y_1])$ ($[X_0, X_1], [Y_0, Y_1]$ is the coordinate in the two-dimensional space), and the function of the prediction query is predicting the approximate number of moving objects within the prediction query region q during a prediction query time bucket t. The prediction process is shown in Fig. 5.19.

In Fig. 5.19, we set the interval between prediction time T_1 and current time T_0 as T which is the same concept in Eq. (5.23). Because of generally discrete collection of historical data in prediction model of time series, this method also uses discretely historical data to achieve the prediction. If prediction query time bucket is $[T_1, T_n]$, the result of each discrete time $T_1, T_2, \ldots, T_n$ of $[T_1, T_n]$ will be added together to get the prediction query answer. And we set the i-th past moment as T_{0-i}. There are two situations for us to discuss:

- When $T_1 = T_n$, it represents the single time, and we may only make the prediction calculation at the single time on the road networks.

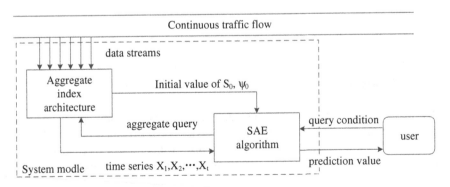

Fig. 5.18 Prediction system model

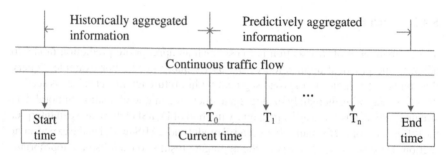

Fig. 5.19 Prediction process

- When $T_1 < T_n$, it represents many discrete times of $[T_1, T_n]$, and we can calculate the sum of prediction value of each time.

  The algorithm which is called *Predict_Agg* is shown as Algorithm 7.

Algorithm 7: Algorithm Predict_Agg(T_0, T_1, T_n)

input : T_0 represents current time,T_1 (T_n) represents prediction start (end) time
1 **begin**
2   Get the aggregate value of current time by aggregate query $\rightarrow S_0$;
3   **if** $T_1 = T_n$ **then**
4   Get the value of Ψ_1 according to $\Psi = \frac{\alpha(t)}{1-(1-\alpha(t))^t}$;
5   According to expression 5.28, Ψ_1 and S_0, get the values of (a_1, b_1, c_1);
6   According to expression 5.23, T_1, T_0 and (a_1, b_1, c_1), calculate Y_1;
7   Return Y_1;
8   **else if** $T_1 < T_n$ **then**
9   for $1 \rightarrow n$, get the values of $\Psi_1, \cdots, \Psi_n$ with $\Psi = \frac{\alpha(t)}{1-(1-\alpha(t))^t}$;
10   for $1 \rightarrow n$, get series of values of $(a_1, b_1, c_1) \rightarrow (a_n, b_n, c_n)$ with expression 5.28, Ψ's value and S_0;
11   for $1 \rightarrow n$, get the values of $Y_1, \cdots, Y_n$ with expression 5.23;
12       $Y = Y_1 + \cdots + Y_n$;
13   Return Y;
14   **end**
15 **end**

5.4.3.2 Evaluation

The road network of Ningbo, China is adapted in the experiment. There were about 1,451 road segments per km^2. There were about 30,760 vehicles and we divided them into four kinds (car, bus, truck and auto-bike) with different speed and moving patterns. The vehicles were uniformly distributed on the road network at the start time.

Spatio-temporal prediction aggregate query based on road networks is based on the aggregation of past and current time to predict results. In this section, we do some experiments according to three factors: (1) historical information length; (2) smoothing parameter α; (3) the length of prediction query time (T, this T is not the same concept as T of Eq. (5.23). And finally, we compare SAES with ES.

When α is set to 0.6, we take an analysis by varying historical information length. When historical information length keeps the value which is bigger than 22, the relative error keeps the lowest. This is because SAES is more dependent on the past time series, and adjusts the smoothing factor based on historical value. Namely, the longer the historical information length is, the smaller the relative error is. When historical information length is long enough, it will have little influence in generating future data. What's more, the longer T is, the bigger the relative error is.

According to the foregoing experiment, we set historical information length to 22, and vary the value of α to find the law. The selection of α's value in prediction model SAES is based on the characteristic of the data distribution, and Ψ dynamically adjust its value depending on the value of α. Here we analyze the error situation whose smoothing parameter α is set in $(0, 1)$. (Historical information length is set to 22) It can be seen that errors in different cycle have different values, when T gets longer, the relative error will be bigger. However, prediction error can be maintained under a certain range with the same value of T, it is due to the SAES model which can dynamically adjust Ψ's value to make it fit.

Finally, we compare our method with ES model in DynSketch. In the DynSketch method, the parameters which make ES model have the best prediction value are: historical information length is set to 4, and smoothing parameter α is set to 0.9. In order to fully reflect the superiority of SAES model, we take those two parameter value to do the experiment, and we get the results.

In summary, we analyzed three factors of SAES in the beginning and got some appropriate values. Then we compared SAES with ES, and found that SAES had superior performance. Our method is a good traffic prediction approach based on aggregated information of data streams in spatio-temporal road networks.

5.4.4 Transition Exponential Smoothing

Studies have shown that the estimates about smoothing parameter must make the prediction error for the previous step be the minimum. Some researchers believe that parameters of the ES model should be able to change over time to meet the latest features of the time series. For example, if the sequence changes with the level (level shifts), the exponentially weighted average must be adjusted such that a larger value corresponds to the latest observations. There are many adaptive methods, such as Holt and Holt-Winters [100] methods, but most have been criticized because of their instability predict. A general method for adaptive smoothing exponential equation is $f_{t+1} = \alpha_t y_t + (1 - \alpha_t)f_t$. f_{t+1} means the predicted value at time $t + 1$. α_t means the prediction smoothing exponential at time t. y_t means the true value at time t. f_t

means the predicted value at time t. Adaptive smoothing methods are available as a collection species smooth transition model, a simple Smooth Transition Regression Model (STR) forms such as: $y_t = a + b_t x_t + e_t$, where $b_t = \omega/(1 + e^{\beta + \gamma V_t})$, α, ω, β, γ are constant parameters. If so, then b_t is a monotonically increasing function of V_t, and the value range is 0 to ω. Smooth Transition Auto Regression Model (STAR) is a similar form, in addition to being replaced calls lag dependencies y_{t-1}.

Taylor [101, 102] made a Smooth Transition Exponential Smoothing (STES) method in 2004, which is a new adaptive exponential smoothing method that smoothing parameter can be modeled as a user-specified variable corresponding mathematical logic function (logistic). This variables simple choice is the size of the prediction error over past time. This method is similar to the method of modeling for the time variation parameters in smooth transition method. This section introduces STES method to traffic data stream aggregate prediction query, and designs data streams aggregate prediction model in road network. STES model prediction equation is as follows:

$$f_{t+1} = \alpha_t y_t + (1 - \alpha_t) f_t, \alpha_t = 1/(1 + e^{\beta + \gamma V_t}) \tag{5.29}$$

The core of smooth transition model is that at least one parameter is modeled as a continuous function of converting variable V_t. As can be seen from Eq. (5.29), the function is limited by α_t to the range from 0 to 1. It is different from the previous model that the historical data is used to estimate α_t, but this is used to estimate the constant parameters β and γ. If $\gamma < 0$, α_t is a monotonically increasing function of V_t. Generally, the choice of V_t is the key to the success of the model. Taking all the adaptive methods into account, the smoothing parameter value depends on changes size of difference about predictive value in the last time. Clearly, errors squared or absolute value of the recent period can be used as conversion variables. Of course, average error, average square error, the percentage error and others can also be used as the conversion variables. Look at STES models prediction equation, and we take $\gamma < 0$. Because when errorV_t expressed becomes large, the description should increase the weight of the true value at time t, namely the weight of y_t. It fits monotonically increasing relationship of prediction equations.

5.4.4.1 Architecture

In using adaptive smooth transition model, we usually hope that the lag prediction error is minimal. It can be seen that when $\gamma = 0$, STES model reduces to ES model. When the STES model chooses historical data, it is targeting to the correct DSD+ structure by Chord platform. This choice of V_t is the absolute error about prediction in the last time for easy to calculate and achieve real-time. β and γ are to be estimated by using historical data through the recursive least two multiplications. When $\gamma = 0$, it is the ES model. When $\gamma < 0$, it can give better predictiing for the data changes in the case of a hierarchy. And when $\gamma > 0$, it is used to compare the reaction about the prediction function to the abnormal value and reduce the impact about the abnormal

Fig. 5.20 STES model
predictions flow

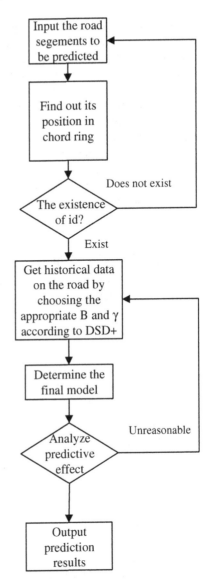

value to the weight of predicting. Figure 5.20 shows the aggregation prediction flow
of STES model based on DSD+ in Chord platform.

STES model has been applied to the inspection about volatility issues in stock
market. It has made more stable results. Its smoothing coefficient is not static, parameter changes can reflect the impact of data for predicting in different periods. The
number of vehicles on road network is also constantly changing volatility. According
to the process of Fig. 5.20, there is the aggregation prediction algorithm after finding
the BPT-tree to the id of the section.

Assuming $qt = (t_0, t_1)$, the algorithm supports two prediction forms. one is $t_0 = t_1$, that is, predicts the next time for a query region, another is $t_1 > t_0$, which predicts the next period of the current time. The *inordernot* algorithm is used to query approximate aggregate values in history time, root is the root node of BPT-tree, x_0, x_1, y_0, y_1, t_0, t_1, represent the x-axis range, the y-axis range, and the time range $t_0 - t_1$, respectively. And n represents the *hislength* using by this algorithm(line 3); $S[i]$ represents the approximate aggregate value get in $time = t - t + t_t$(line 6); *error* represents the prediction error absolute in $time = t - t + t_t$(line 7); α_t represents the smoothing parameter in $time = t - t + t_t$, V_t represents the absolute value of the simple calculation error(line 8); S_{next} represents the predictive value in $time = t + 1 - t + t_t + 1$(line 9). The non-recursive in-order traversal BPT-tree in this method gets the approximate aggregate values based on sketch information within the query range.

Algorithm 8: agg_prediction($x_0, x_1, y_0, y_1, t_0, t_1$)

---

 input : query region: (x_0, x_1, y_0, y_1) and predictive time range: (t_0, t_1)
 output: aggregation results
1 **begin**
2     $t_t = t_1 - t_0$;
3     $S_{next} = S[t_0 - n - t_t] = inordernot(root, x_0, x_1, y_0, y_1, t_0 - n - t_t, t_0 - n)$;
4     $t = t_0 - n - t_t$;
5   **for** $i = t$ *to* $t_0 - 1$ **do**
6         $S[i] = inordernot(root, x_0, x_1, y_0, y_1, i, i + t_t)$;
7         $error = fabs(S_{next} - S[i])$;
8         $\alpha_t = 1/(1 + exp(\beta + r \cdot error))$;
9         $S_{next} = \alpha_t \cdot S[i] + (1 - \alpha_t) \cdot S_{next}$;
10   **end**
11   **return** S_{next};
12 **end**

5.5 Summary

Precise query and aggregate query are two typical kinds of queries in road network. Precise queries which would get exact location in road network are used widely in ITS. And aggregate query aims at obtaining summarized information such as vehicles counts. Some queries mentioned above refer to precise query, such as nearest neighbor query, continuous nearest neighbor query and reverse search method of CNN based on Cyclic Optimal Multi-step Algorithm. Other queries mentioned above refer to aggregate query, such as ES, SAES, MH and STES methods.

In the next chapter we will introduce the development status of the national intelligent transportation technology and new trend of intelligent traffic under the background of cloud computing and big data.

Chapter 6
The Trend of Development

Since the 1960s, intelligent transportation technologies have been proposed with its rapid development, it has been widely used in various countries, effectively easing road congestion and improving travel efficiency, which has achieved great social and economic benefits. However, with the rapid economic development of the world, number of vehicles is increasing, and the traffic problem is getting worse. In this case, intelligent traffic is imminent and modern intelligent transportation management system is coming into being. The current transportation systems generally use special equipment and build on the top of the dedicated systems with high cost. With the rapid development of modern transportation, the data which is based on the road networks is growing, which makes the transportation system more and more difficult to cope with. How to store, process, analyze, mine and utilize massive traffic information has gradually become a bottleneck restricting the development of intelligent transportation.

Cloud computing technology is a new type of computing patterns, which embodies a new concept of information services. Cloud computing is the key technique of solving the problem of massive data with its automated computer resource scheduling, deployment of high-speed information and excellent scalability. As an emerging computing and business model, cloud computing accelerates the processes of transportation information service and information industry. The rapid development of cloud computing in the field of intelligent transportation applications has positive significance to improve the integrated information processing capacity of the cities and promotes the upgrading of the industrial optimization and the structuring. At the same time, cloud computing promotes the transformation of economic development mode, which has a broad market prospect.

In this chapter, we will introduce the systems and applications of ITS based on cloud computing and big data.

J. Feng and T. Watanabe, *Index and Query Methods in Road Networks*,
Smart Innovation, Systems and Technologies 29,
DOI 10.1007/978-3-319-10789-9_6

6.1 Intelligent Transportation Cloud

There are several sub-systems of ITS, and these subsystems' computing devices and application services can be specified to obtain general-purpose computing device layer which can be further used in intelligent transportation systems construction. This general-purpose computing device layer can use the provider services of the current cloud computing service providers, so that the services provided by the ITS will become applications of cloud computing services. This kind of ITS which is based on the services of cloud computing is called "intelligent transportation cloud". Intelligent transportation cloud's computing and storage capacity will not be restricted, and can realize the exchange of traffic information to provide users infrastructure, computing platform and basic traffic data.

As shown in Fig. 6.1, it is typical intelligent transportation cloud structure which has a virtual layer on the hardware device layer (X86 servers, Unix 6 servers), and the virtual layer provides virtual machines (Unix virtual machines, Linux virtual machines, Windows virtual machines) on the basis of the hardware resource layer and virtualization layer (storage virtualization SVC and server virtualization VMWare/Xen/PowerVM), which is composed of a virtual machine application system. And these three layers combine with cloud service management to form IaaS (Infrastructure as a Service) layer. Cloud computing services management is independent of the virtual machines and the virtualization layer, and it provides service

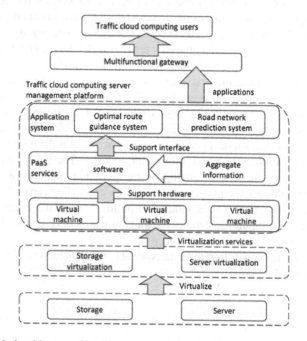

Fig. 6.1 Logical architecture of intelligent transportation cloud

portal, service catalog, unified monitoring, resource management and other functions. It can construct PaaS (Platform as a service) layer on the basis of IaaS, and the PaaS provides the mainstream software products with self-installation and deployment, including applications to provide aggregate information for the upper index interface. The corresponding application systems such as the traffic management, the network prediction system, the optimal route guidance system, simulation decision support system can run on PaaS layer or run directly on IaaS layer.

Cloud computing provides a new platform in handling traffic business applications. Different intelligent transportation business applications are deployed at different levels of cloud platform. The corresponding application systems such as traffic management, the network prediction system, the optimal route guidance system, the simulation decision support system can run on PaaS layer or run directly on IaaS layer. Optimal path navigation services deployed on PaaS layer, for example, through the use of PaaS cloud, integrate vehicles, people, roads and other comprehensive traffic factors based on cloud computing data center. By processing and analysis, and information integration, information to the road users is published through the electronic map, car terminals, real-time SMS and broadcast media, providing them with the optimal route guidance information and a variety of real-time traffic information service to help change route in advance for drivers so as to avoid traffic congestion and improve traffic efficiency and safety.

However, cloud computing technology which provides a new platform for intelligent transportation systems also brings new technical challenges. The computing environments of the traditional application of intelligent transportation business, depend on the particular computing environment, present strict requirements for users and applications. The user must write and run the program according to the environment provided and it is almost impossible to transplant programs to new architectures or operating systems, or it must spend a lot of manpower and time. If it cannot be compatible with existing business applications, the software will not only result in a waste of resources, and even it is difficult to read and analyze the valuable data. Therefore, we need to study how to build extensive and compatible computing environment with virtualization technology, and achieve the "zero modification" cloud of migration for the traditional intelligent transportation business applications.

Cloud computing platform is a typical distributed computing environment, and its powerful MapReduce parallel processing mode provides the possibility of greatly improving the efficiency of data processing and querying, but also presents new technical challenges on the traditional method of indexing and query (such as restricted in the R-Tree spatial overlap division, R-TPR$^\pm$ Tree, MOR-tree and Sketch RR-Tree, etc. based on R-Tree index structure will limit the MapReduce parallel ability to fully play). How to build traditional road network data processing method and interface of parallel processing model with cloud platform, and how to achieve a parallel transformation of traditional query and processing methods will be technical issues of the intelligent transportation cloud urgent to study. In addition, as a typical cloud platform, HDFS cannot support random write for files and it is also a new challenge to efficient storage for frequent update traffic flow data.

Intelligent transportation cloud provides for the transportation business process data processing and service platform with high availability, high reliability and high fault tolerance, however, with the development of ITS, more and more mobile devices are used in traffic information collection which forms a large amount of traffic data. For instance, in the ITS of a big city, supposing that there are one million GPS enabled vehicles, the vehicles emit one record every 30 or 60 s, each record contains ID, CompanyID, VehicleSimID, GPSTime, GPSLongitude, GPSLatitude and other related attributes, each reord is 100 Bytes, then the total size of the data per day will be $100B * 10^6/\text{min} * 60/\text{h} * 24/\text{day} \approx 144G$ [103]. The massive traffic data brings new challenges to the department of transportation since these data comes into being transportation big data which is difficult to store and query with using most relational database management systems and desktop statistics and visualization packages, requiring instead of "massively parallel software running on tens, hundreds, or even thousands of servers".

6.2 The Storage Techniques for Transportation Big Data

The relational database management systems have been empowered with rich functionality by using K-d tree and R-tree indexes to support efficient multi-dimensional access, but RDBMS cannot scale up well to deal with huge volume data and support millions of insert operations per minute [103]. As the data grows in traffic, new techniques and approaches need to be adopted. Literature [104] proposes the TrajStore which is a storage system designed to segment trajectories and co-locate trajectory segments that are geographically and temporally near each other. It slices trajectories into subtrajectories that fit into spatio-temporal regions, and densepacks the data about each region in a block (or collection of blocks) on disk. TrajStore uses an adaptive multi-levels grid [105] over those blocks to look up data in space and a sparse index in time to answer historical queries (which can be formulated as hypercubes).

With the continuous development of cloud computing technology, the Key-value stores, such as BigTable [106], Hbase [107] can support millions of updates per minute while providing fault tolerance and high availability. HBase is an open source, non-relational, distributed database modeled after Google's BigTable and is written in Java. It was developed as a part of Apache Software Foundation's Apache Hadoop project and runs on top of HDFS (Hadoop Distributed Filesystem), providing BigTable-like capabilities for Hadoop. That is, it provides a fault-tolerant way of storing large quantities of sparse data (small amounts of information caught within a large collection of empty or unimportant data, such as finding the 50 largest items in a group of 2 billion records, or finding the non-zero items representing less than 1/10 of 1 % of a huge collection).

In HBase, tables are partitioned horizontally into regions, which are the units that get distributed over the HBase cluster. HBase infrastructure architecture is also based on a distributed master-slave architecture. The HBase master node orchestrates

Fig. 6.2 Hbase NoSQL
database system
architecture [108]

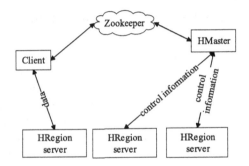

a cluster of one or more slaves. HBase also depends on a quorum service. HBase maintains all its data via Hadoop filesystem APIs. Figure 6.2 shows an HBase system architecture. The clients connect to query service to find the location of the metadata, which in turn points to the region with data tables. This is the operation for the first interaction; thereafter, the client interacts directly with the hosting region server in the cluster. Clients also cache their learning and continue to use the cached entries until they fail. The writes are appended to the commit log in the region server HDFS and then added to an in-memory memstore. The memstore is flushed to the file system eventually.

Making full use of HBase's features such as high insert throughput and large data volumes, fault-tolerance, and high availability to solve storage problems, Literature [108, 109] has made researches on Hbase-based intelligent big data storage methods. Although Key-value store model supports millions of updates per minute while providing fault tolerance and high availability, they can not provide rich functionality and do not support multi-dimensional access natively. They can support efficient point and range queries on rowkey, but it has to scan the whole table for the queries on non-rowkeys. Although the MapReduce framework can be used to enhance concurrency and improve the query performance, full scan is still wasteful, especially for the high selective query.

In order to support both high insert throughput and efficient multi-dimensional query,we need to create multi-index for data, QT-Chord [110], RT-CAN [111], EMINC [112], Literature [113] and Literature [103] have made researches on Hadoop-based indexing methods from different perspectives. However, these methods mainly concerned Euclidean-space data, and there are still many challenges to index traffic big data, such as: modeling in distributed environment and distinct counting problem. Literature [114] proposed the DynSketch method which can improve the accuracy of the results through adjusting the number of sketches. However, it also brings a new problem: operations between different number of sketches would bring great relative errors, especially in a distributed environment. The accumulative errors of different local nodes would reduce overall accuracy. In addition, different number of local sketches cannot execute operation "OR", which would cause no global solution.

6.3 Challenges to Transportation Big Data Processing

The "variety" of data types is one characteristic of "3V" of big data. Traffic data has wide sources and they cannot be represented by a simple data structure. Computer is good at dealing with homogeneous data, except can handled heterogeneous data efficiently. How the data are organized into a reasonable structure is an important issue. As traffic data can be obtained in multi-ways, the data often contain incomplete information and false data. Incompleteness of the data must be addressed at analysis stage effectively. The approach is a challenge and recent researches which concern probabilistic data management will provide new methods to deal with uncertain and incomplete data.

Today, the rapid expansion of the size of data is far beyond current computer processing power. Turing Award winner Jim Gray and IDC company have predicted that the global amount of data doubles every 18 months. The current global data storage and processing capacity has lagged far behind the growth rate of the data. For example, in Shanghai and Shenzhen GPS data contain around 2 billion records everyday, and increase at a rate of 1–2 GB/day. For big data, the data processing speed is very important. In general, the larger the data analysis is, the longer processing time will be. If we design a certain system to deal with specific data, its processing speed may be very fast, but it does not meet the general requirements of big data. In many cases, users want to obtain results of the data analysis immediately, for example, a traffic management system needs to find all the traffic jams on the travel route and give information on alternate travel routes. At this time, the system would execute nearby query on the route of moving objects. When the data size is growing, it is a challenge to develop efficient query processing algorithms. Currently, the efficiency of big data processing has become a central issue and the data processings in different stages are different. Traditional mathematical methods have been unable to adapt to uncertain, dynamic analysis of big data. We need to combine computational science, mathematics, physics and other disciplines to create a new scientific method of data so that we can research data patterns and statistical characteristics on the premise that data have the diversity and uncertainty features.

According to the volume and distribute features, traditional methods are not suitable for processing massive data. So we need new methods to process massive data concurrently and there comes a series of work on MapReduce. MapReduce is a model proposed by Google to process and generate big data. Hadoop is the open-source realization of MapReduce and the big data processing technologies concerned by the academic and private sectors. As the parallel programming model is easy to use, there are many big data processing query language, such as Pig of Yahoo and Sawzall of Google. These languages will parse query into a series of MapReduce jobs and execute them on the distributed file systems. Compared with MapReduce, high level query languages are suitable for users to process massive data. In academia, literatures [115–117] have researched k-NN and top-k join with MapReduce. However, they have shortages on real-time and efficiency. For distributed data, parallel computation process Shuffle would send all key-value pairs whose key is the same in intermediate stages results to the same target node through the network. This will bring additional data movement costs, each node needs to share network bandwidth and network bandwidth is the most valuable resource in the large clusters and data centers. How to reduce the intermediate result processing costs of MapReduce and how to eliminate data movement bottleneck are the technical problems in the intelligent transportation big data processing and even general big data processing fields.

6.4 Knowledge Discovery from Transportation Big Data

"Big Data", characteristic of the times, will continue to be assets for institutions, as a powerful weapon for institutions and companies to enhance the competitiveness, and the full use of large data which contain rich value will bring industries and enterprises with strong competitiveness. McKinsey Global research institutions indicated in the "big data: innovation, competition and productivity in the next frontier" in May 2011 that taking full advantage of big data can help global personal positioning ISP increase $100 billion revenues, help the European public sector enhance an annual output value of $250 billion of the management, help the U.S. healthcare industry increase output value of $300 billion annually, and help the United States to obtain over 60 % of retail net profit growth. In the field of intelligent transportation, with GPS navigation systems and location-aware devices widely used, a rich mobile trajectory data has formed. Through analysis data mining for these valuable resources, we will discover the natural laws and human activities information, and thus provide new ideas to address the growing severe problems of urban traffic. The mobile trajectory data analysis mainly researches how to extract authentic trajectory of moving objects from massively noised raw data, and then analyzes features of the trajectory of the vehicle or the crowded. Based on these, we study the relationships with city road topology, the driving habits or personal behavior, as well as the applications of the traffic tracing characteristics in transportation and planning fields. There are already a large number of scholars doing analysis and mining research of mobile trajectory data from different angles: John Krumm et al., who used the vehicle trajectory data to recover vehicle road information, analyzed and calculated the width of the road, the traffic direction and the road crossing conditions, to do a series of work in the city reconstruction of a road map [118, 119]. Liu et al., who used taxi traveling speed information, indentified the road hot spots [120]. Rajesh Krishna Balan et al., who analyzed the taxi trajectory data to help people before taking a taxi have a clear understanding of both the expected travel time and costs [121]. As the complexity of traffic conditions, traffic speed is not solely dependent on the distance between the two places. In order to help the driver find the optimal route, Xie and Zheng et al., who studied with the taxi trajectory data, using the rich taxi driver driving experience, provided people with navigation recommendations [122, 123], Li et al., studied how to predict the number of passengers for a taxi based on historical data to help drivers quickly find passengers [124]; Zhang et al., analyzed anomalous driving patterns from taxi's GPS traces [125].

Vehicle, as the crowd moving carrier, its trajectory contains a wealth of information on population movement. Based on the analysis of trajectory data, you can further excavate the move law of the crowds. The main research is how to mine crowd moving information from trajectory data, such as characteristic quantity of flow and density, and thus how to get the crowd moving transformation law of spatial and temporal environments from mining analysis, such as cyclical, scaleless and other universal laws, simultaneously, and analyze regional social functions, land use and other macro factors on the impact of population movement law. Zhejiang University, having extracted sequential variation of the number of moving people from the taxi trajectory data, found these sequential data may reflect the social function of the corresponding region of cities [126]. The Asia Microsoft Research, from the perspective of transportation planning, analyzed detour, low speed, etc. due to traffic planning from taxi trajectory data, and detected the current transportation planning problems [127].

6.5 Summary

In order to deal with massive traffic data in applications, cloud and big data have been introduced into road network techniques to support parallel processing. Cloud computing provides a new platform in handling traffic applications with high availability, high reliability and high fault tolerance. However, more and more mobile devices are used in traffic information collection which forms a large amount of traffic data, but the rapid expansion of data size is far beyond current computer processing power. Process and storage for big data are key problems to deal with. In addition, knowledge discovery from big data which contain rich value can help us obtain more information and will bring industries and enterprises with strong competitiveness.

References

1. O. Wolfson, S. Chamberlain, S. Dao, L.Q. Jiang, Location management in moving objects databases. in *Proceedings of Workshop on Satellite-Based Information Systems*. (1997), pp. 7–13
2. D. Pfoser, Indexing the trajectories of moving objects. IEEE Data Eng. Bull. **25**, 3–9 (2002)
3. J.L. Bentley, Multidimensional binary search trees used for associative searching. Commun. ACM **18**(9), 509–517 (1975)
4. J.L. Bentley, Multidimensional binary search trees in database applications. IEEE Trans. Softw. Eng. SE-5(4), 333–340 (1979)
5. J.T. Robinson, The k-d-b-tree: a search structure for large multidimensional dynamic indexes. in *Proceedings of the 1981 ACM SIGMOD international conference on Management of Data, SIGMOD '81* (ACM, New York, 1981), pp. 10–18
6. R. Bayer, E. McCreight, Organization and maintenance of large ordered indices. in *Proceedings of the 1970 ACM SIGFIDET (now SIGMOD) Workshop on Data Description, Access and Control, SIGFIDET '70* (ACM, New York, 1970), pp. 107–141
7. D. Comer, Ubiquitous b-tree. ACM Comput. Surv. **11**(2), 121–137 (1979)
8. H. Fuchs, G.D. Abram, E.D. Grant, in *Siggraph '83: Near real-time shaded display of rigid objects*, ed. by P.P. Tanner,in *Proceedings of the 10th annual conference on Computer graphics and interactive techniques*, (ACM, New York, 1983), pp. 65–72
9. H. Fuchs, Z.M. Kedem, B.F. Naylor, On visible surface generation by a priori tree structures. SIGGRAPH Comput. Graph. **14**(3), 124–133 (1980)
10. T. Matsuyama, L.V. Hao, M. Nagao, A file organization for geographic information systems based on spatial proximity. Comput. Vis. Graph. Image Process. **26**(3), 303–318 (1984)
11. J.B. Rosenberg, Geographical data structures compared: a study of data structures supporting region queries. Trans. Comput. Aided Des. Integr. Circ. Syst. **4**(1), 53–67 (2006)
12. J. Banerjee, W. Kim, Supporting vlsi geometry operations in a database system. in *Proceedings of the Second International Conference on Data Engineering* (IEEE Computer Society, Washington, 1986), pp. 409–415
13. B.C. Ooi, Spatial kd-tree: A data structure for geographic database. in *BTW* (1987), pp. 247–258
14. B.C. Ooi, R.S. Davis, K.J. Mcdonnell, Extending a dbms for geographic applications. in *Proceedings of The 5th International Conference on Data Engineering* (IEEE Computer Society, Washington, 1989), pp. 590–597
15. wikipedia. B-tree. http://en.wikipedia.org/wiki/B-tree (2013)
16. D. Knuth, Sorting and searching. in The Art of Computer Programming, vol. 3 (Addison-Wesley Professional, 1998)

© Springer International Publishing Switzerland 2015

J. Feng and T. Watanabe, *Index and Query Methods in Road Networks*,
Smart Innovation, Systems and Technologies 29,
DOI 10.1007/978-3-319-10789-9

17. A. Guttman, R-trees: a dynamic index structure for spatial searching. in *Proceedings of the 1984 ACM SIGMOD International Conference on Management of Data, SIGMOD '84* (ACM, New York, 1984), pp. 47–57

18. N. Beckmann, H.P. Kriegel, R. Schneider, B. Seeger, The r*-tree: an efficient and robust access method for points and rectangles. in *Proceedings of the 1990 ACM SIGMOD International Conference on Management of Data, SIGMOD '90* (ACM, New York, 1990), pp. 322–331

19. T.K. Sellis, N. Roussopoulos, C. Faloutsos, The r+-tree: A dynamic index for multi-dimensional objects. in *Proceedings of the 13th International Conference on Very Large Data Bases, VLDB '87* (Morgan Kaufmann Publishers Inc., San Francisco, 1987), pp. 507–518

20. I. Kamel, C. Faloutsos, Hilbert r-tree: An improved r-tree using fractals. in *Proceedings of the 20th International Conference on Very Large Data Bases, VLDB '94* (Morgan Kaufmann Publishers Inc., San Francisco, 1994), pp. 500–509

21. H.V. Jagadish, Spatial search with polyhedra. in *Proceedings of the Sixth International Conference on Data Engineering* (1990), pp. 311–319

22. M. Schiwietz, Speicherung und Anfragebearbeitung komplexer Geo-Objekte (PhD Thesis, Ludwig-Maximilian-Universitat Munchen, 1993).

23. R.A. Finkel, J.L. Bentley, Quad-trees; a data structure for retrieval on composite keys. Acta Informatica **4**(1), 1–9 (1974)

24. G. Kedem, The quad-cif tree: a data structure for hierarchical on-line algorithms. in *Proceedings of 19th Conference on Design Automation* (IEEE, 1982), pp. 352–357

25. R. Fagin, J. Nievergelt, N. Pippenger, H.R. Strong, Extendible hashing a fast access method for dynamic files. ACM Trans. Database Syst. **4**(3), 315–344 (1979)

26. H.P. Kriegel, B. Seeger, Multidimensional order preserving linear hashing with partial expansions. in *Proceedings of the International Conference on Database Theory, ICDT '86* (Springer, London, 1986), pp. 203–220

27. P.A. Larson, Dynamic hashing. Bit. Numer. Math. **18**(2), 184–201 (1978)

28. K. Hinrichs, J. Nievergelt, The grid file: A data structure to support proximity queries on spatial objects. in *Workshop on Graph-Theoretic Concepts in Computer*, Science (1983), pp. 100–113

29. J. Nievergelt, H. Hinterberger, K.C. Sevcik, The grid file: An adaptable, symmetric multi-key file structure, in *Trends in Information Processing Systems, vol. 123*, Lecture Notes in Computer Science, ed. by Arie Duijvestijn, PeterChristian Lockemann (Springer, Berlin, 1981), pp. 236–251

30. J. Nievergelt, H. Hinterberger, K.C. Sevcik, The grid file: an adaptable, symmetric multikey file structure. ACM Trans. Database Syst. **9**(1), 38–71 (1984)

31. M. Tamminen, Efficient spatial access to a data base. in *Proceedings of the 1982 ACM SIGMOD International Conference on Management of Data, SIGMOD' 82* (ACM, New York, 1982), pp. 200–206

32. M. Tamminen, The extendible cell method for closest point problems. BIT Numer. Math. **22**(1), 27–41 (1982)

33. A. Hutflesz, H.-W. Six, P. Widmayer, The r-file: an efficient access structure for proximity queries. in *Proceedings of the Sixth International Conference on Data Engineering* (1990), pp. 372–379

34. H.W. Six, P. Widmayer, Spatial searching in geometric databases. in *Proceedings of the Fourth International Conference on Data Engineering* (IEEE Computer Society, Washington, 1988), pp. 496–503

35. J.A. Orenstein, Spatial query processing in an object-oriented database system. in *Proceedings of the 1986 ACM SIGMOD International Conference on Management of Data, SIGMOD '86* (ACM, New York, 1986), pp. 326–336

36. G.M. Morton, A computer oriented geodetic data base and a new technique in file sequencing. in *IBM Germany Scientific Symposium Series* (1966)

37. D. Hilbert, Uber die stetige abbildung einer linie auf ein flachenstck. in Gesammelte Abhandlungen (Springer, Berlin, 1970), pp. 1–2

38. G. Peano, Sur une courbe, qui remplit toute une aire plane. Math. Ann. **36**(1), 157–160 (1890)

39. Y. Leung, K.S. Leung, J.Z. He, A generic concept-based object-oriented geographical information system. Int. J. Geograph. Inf. Sci. **13**(5), 475–498 (1999)
40. A. Frank, S. Simpf, Multiple representations for cartographic objects in a multi-scale tree: an intelligent graphical zoom. Comput. Graph. **18**(6), 823–829 (1994)
41. S. Timpf, Hierarchical structures in map series. http://www.geoinfo.tuwien.ac.at/publications/formerPersonnel/timpf/diss/table-of-contents.htm (1998)
42. S. Spaccapietra, C. Parent, E. Zimanyi, Murmur: a research agenda on multiple representations. in *Proceedings of the 1999 International Symposium on Database Applications in Non-Traditional Environments, DANTE '99* (IEEE Computer Society, Washington, 1999), pp. 373–384
43. C. Parent, S. Spaccapietra, E. Zimany, Spatio-temporal conceptual models: data structures + space + time. in *Proceedings of the 7th ACM international symposium on Advances in Geographic Information Systems, GIS '99* (ACM, New York, 1999), pp. 26–33
44. C. Parent, S. Spaccapietra, E. Zimany, P. Donini, C. Plazanet, C. Vangenot, Modeling spatial data in the mads conceptual model. in *Proceedings of the 8th International Symposium on Spatial Data Handling, SDH'98* (1998), pp. 138–150
45. C. Parent, S. Spaccapietra, E. Zimanyi, Murmur: database management of multiple representations. in *Proceedings of AAAI-2000 Workshop on Spatial and Temporal Granularity* (Austin, Texas, 2000)
46. K. Horikawa, M. Arikawa, H. Takakura, Y. Kambayashi, Dynamic map synthesis utilizing extended thesauruses and reuse of query generation process. in *Proceedings of the 5th ACM International Workshop on Advances in Geographic Information Systems, GIS '97* (ACM, New York, 1997), pp. 9–14
47. E. Keighan, Managing spatial data within the framework of the relational model. in *Technical Report* (Oracle Corporation, Canada, 1993)
48. M.J. Egenhofer, Deriving the composition of binary topological relations. J. Visiual Languanges Comput. **5**, 133–149 (1994)
49. Geospatial Information Authority of Japan. Map 2500. http://www.gsi.go.jp/MAP/CD-ROM/2500/t2500.htm
50. S. Shekhar, D.R. Liu, Ccam: A connectivity-clustered access method for aggregate queries on transportation networks: a summary of results. in *Proceedings of the Eleventh International Conference on Data Engineering, ICDE '95* (IEEE Computer Society, Washington, 1995), pp. 410–419
51. M.F. Goodchild, Gis and transportation: Status and challenges. GeoInformatica **4**(2), 127–139 (2000)
52. S. Winter, Modeling costs of turns in route planning. GeoInformatica **4**, 345–361 (2002)
53. J. Fawcett, P. Robinson, Adaptive routing for road traffic. IEEE Comput. Graph. Appl. **20**(3), 46–53 (2000)
54. C.Y. Lee, An algorithm for path connections and its applications. IEEE Trans. Electron. Comput. **10**(3), 346–365 (1961)
55. D. Papadias, J. Zhang, N. Mamoulis, Y. Tao, Query processing in spatial network databases. in *Proceedings of the 29th International Conference on Very Large Data Bases*, vol. 29 (VLDB Endowment, 2003), pp. 802–813
56. N. Christofides, *Graph Theory : An Algorithmic Approach* (Academic Press Incorporated, Orlando, 1975)
57. J. Chung, O. Peak, J. Lee, K.H. Ryu, in *Temporal pattern mining of moving objects for location-based service*, ed. by A. Hameurlain, R. Cicchetti, L.R. Traunm, Database and Expert Systems Applications, (Springer, Heidelberg, 1976), pp. 331–340
58. M. Hadjieleftheriou, G. Kollios, V. Tsotras, D. Gunopulos, Efficient indexing of spatiotemporal objects. in *Proceedings of the 8th International Conference on Extending Database Technology: Advances in Database Technology, EDBT '02* (Springer, London, 2002), pp. 251–268
59. G. Kollios, D. Gunopulos, V. Tsotras, A. Delis, M. Hadjieleftheriou, Indexing animated objects using spatiotemporal access methods. IEEE Trans. Knowl. Data Eng. **13**(5), 758–777 (2001)

60. I. Lazaridis, K. Porkaew, S. Mehrotra, Dynamic queries over mobile objects. in *Proceedings of the 8th International Conference on Extending Database Technology: Advances in Database Technology, EDBT '02* (Springer, London, 2002), pp. 269–286

61. D. Pfoser, C.S. Jensen, Indexing of network constrained moving objects. in *Proceedings of the 11th ACM International Symposium on Advances in Geographic Information Systems, GIS '03* (ACM, New York, 2003), pp. 25–32

62. P.K. Agarwal, L. Arge, J. Erickson, Indexing moving points. J. Comput. Syst. Sci. **66**(1), 207–243 (2003)

63. S. Saltenis, C.S. Jensen, Indexing of moving objects for location-based services. in *Proceedings of the 18th International Conference on Data Engineering, ICDE '02* (IEEE Computer Society, Washington, 2002), pp. 463–472

64. S. Saltenis, C.S. Jensen, S. Leutenegger, M. Lopez, Indexing the positions of continuously moving objects. in *Proceedings of the 2000 ACM SIGMOD International Conference on Management of Data, SIGMOD '00* (ACM, New York, 2000), pp. 331–342

65. A.P. Sistla, O. Wolfson, S. Chamberlain, S. Dao, Modeling and querying moving objects. in *Proceedings of the Thirteenth International Conference on Data Engineering, ICDE '97* (IEEE Computer Society, Washington, 1997), pp. 422–432

66. A. Civilis, C.S. Jensen, S. Pakalnis, Techniques for efficient road-network-based tracking of moving objects. IEEE Trans. Knowl. Data Eng. **17**(5), 698–712 (2005)

67. D. Papadias, Y. Tao, P. Kalnis, J. Zhang, Indexing spatio-temporal data warehouses. in *Proceedings of the 18th International Conference on Data Engineering 2002* (2002), p. 166C175

68. Y.F. Tao, G. Kollios, J. Considine, F.F. Li, D. Papadias, Spatio-temporal aggregation using sketches. in *Proceedings of the 20th International Conference on Data Engineering, ICDE '04* (IEEE Computer Society, Washington, 2004), pp. 214–226

69. J. Sun, D. Papadias, Y. Tao, B. Liu, Querying about the past, the present, and the future in spatio-temporal databases. in *Proceedings of the 20th International Conference on Data Engineering, ICDE '04* (IEEE Computer Society, Washington, 2004), pp. 202–213

70. Y.L. Zhu, X. Ren, J. Feng, Nco-tree: a spatio-temporal access method for segment-based tracking of moving objects. in *Proceedings of the 10th International Conference on Knowledge-Based Intelligent Information and Engineering Systems-Volume Part II, KES'06* (Springer, Berlin, 2006), pp. 1191–1198

71. P. Flajolet, G. Martin, Probabilistic counting algorithms for data base applications. J. Comput. Syst. Sci. **32**(2), 182–209 (1985)

72. B.Y. Tan, N. Liu, Technology of approximate query over data stream. WORLD SCI-TECH R and D **28**(2), 57–60 (2006)

73. H.V. Jagadish, N. Koudas, S. Muthukrishnan, V. Poosala, K. Sevcik, T. Suel, Optimal histograms with quality guarantees. in *Proceedings of the 24rd International Conference on Very Large Data Bases, VLDB '98* (Morgan Kaufmann Publishers Inc., San Francisco, 1998), pp. 275–286

74. V. Poosala, P.J. Haas, Y.E. Ioannidis, E.J. Shekita, Improved histograms for selectivity estimation of range predicates. in *Proceedings of the 1996 ACM SIGMOD International Conference on Management of Data, SIGMOD '96* (ACM, New York, 1996), pp. 294–305

75. N. Thaper, S. Guha, P Indyk, N. Koudas, Dynamic multidimensional histograms. in *Proceedings of the 2002 ACM SIGMOD International Conference on Management of Data, SIGMOD '02* (ACM, New York, 2002), pp. 428–439

76. M.F. Mokbel, T.M. Ghanem, W.G. Aref, Spatio-temporal access methods. IEEE Data Eng. Bull. **26**, 40–49 (2003)

77. S. Šaltenis, C.S. Jensen, S.T. Leutenegger, M.A. Lopez, Indexing the positions of continuously moving objects. in *Proceedings of the 2000 ACM SIGMOD International Conference on Management of Data, SIGMOD '00* (ACM, New York, 2000), pp. 331–342

78. D. Lin, C.S. Jensen, B.C. Ooi, S. Saltenis, Efficient indexing of the historical, present, and future positions of moving objects. in *Proceedings of the 6th International Conference on Mobile Data Management, MDM '05* (ACM, New York, 2005), pp. 59–66

79. C.Q. Jin, W.B. Guo, F.T. Zhao, Getting qualified answers for aggregate queries in spatio-temporal databases. in *Proceedings of the joint 9th Asia-Pacific Web and 8th International Conference on Web-Age Information Management Conference on Advances in Data and Web Management, APWeb/WAIM '07* (Springer, Berlin, 2007), pp. 220–227

80. F. Rusu, A. Dobra, Statistical analysis of sketch estimators. in *Proceedings of the 2007 ACM SIGMOD International Conference on Management of Data, SIGMOD '07* (ACM, New York, 2007), pp. 187–198

81. Y. Tao, X. Lian, D. Papadias, M. Hadjieleftheriou, Random sampling for continuous streams with arbitrary updates. IEEE Trans. Knowl. Data Eng. **19**(1), 96–110 (2007)

82. S. Chaudhuri, G. Das, M. Datar, R. Motwani, V. Narasayya, Overcoming limitations of sampling for aggregation queries. in *Proceedings of the IEEE International Conference on Data Engineering* (2001), pp. 534–542

83. J. Feng, C.Y. Lu, Research on novel method for forecasting aggregate queries over data streams in road networks*. J. Front. Comput. Sci. Technol. **4**(11), 1027 (2010)

84. T. Brinkhoff, H.P. Kriegel, R. Schneider, B. Seeger, Multi-step processing of spatial joins. in *Proceedings of the 1994 ACM SIGMOD International Conference on Management of Data, SIGMOD '94* (ACM, New York, 1994), pp. 197–208

85. A. Henrich, A distance scan algorithm for spatial access structures. in *Proceeding of the Second ACM Workshop on Geographical Information Science* (Citeseer, 1994), pp. 136–143

86. S. Bespamyatnikh, J. Snoeyink, Queries with segments in voronoi diagrams. in *Proceedings of the Tenth Annual ACM-SIAM Symposium on Discrete Algorithms, SODA '99* (Society for Industrial and Applied Mathematics, Philadelphia, 1999), pp. 122–129

87. Z.X. Song, N. Roussopoulos, K-nearest neighbor search for moving query point. in *Proceedings of the 7th International Symposium on Advances in Spatial and Temporal Databases, SSTD '01* (Springer, London, 2001), pp. 79–96

88. Y.F. Tao, D. Papadias, Q.M. Shen, Continuous nearest neighbor search. in *Proceedings of the 28th International Conference on Very Large Data Bases, VLDB '02* (VLDB Endowment, 2002), pp. 287–298

89. E.P.F. Chan, N. Zhang, Finding shortest paths in large network systems. in *Proceedings of the 9th ACM International Symposium on Advances in Geographic Information Systems, GIS '01* (ACM, New York, 2001), pp. 160–166

90. Y.W. Huang, N. Jing, E.A. Rundensteiner, Path queries for transportation networks: dynamic reordering and sliding window paging techniques. in *Proceedings of the 4th ACM International Workshop on Advances in Geographic Information Systems, GIS '96* (ACM, New York, 1996), pp. 9–16

91. Y.W. Huang, N. Jing, E.A. Rundensteiner, Effective graph clustering for path queries in digital map databases. in *Proceedings of the Fifth International Conference on Information and Knowledge Management, CIKM '96* (ACM, New York, 1996), pp. 215–222

92. N. Jing, Y.W. Huang, E.A. Rundenstteiner, Hierarchical encoded path views for path query processing:an optimal model and its performance evaluation. IEEE Trans. Knowl. Data Eng. **10**(3), 409–431 (1998)

93. C. Han, S. Song, C.H. Wang, A real-time short-term traffic flow adaptive forecasting method based on arima model. J. Syst. Simul. **16**(7), 1530–1535 (2004)

94. J.Z. Li, L.J. Guo, D.D. Zhang, W.P. Wang, Processing algorithms for predictive aggregate queries over data streams. J. Softw. **16**(7), 1252–1261 (2005)

95. M. Ben, E. Cascetta, Recent progress in short-range traffic prediction, compendium of technical papers. Inst. Transp. Eng. (ITE), 262–265 (1993)

96. J. Feng, Z.H. Zhu, Y.Q. Shi, L.M. Xu, A new spatioctemporal prediction approach based on aggregate queries. Int. J. Knowl. Web Intell. **4**(1), 20–33 (2013)

97. J. Feng, Z.H. Zhu, R.W. Xu, A traffic flow prediction approach based on aggregated information of spatio-temporal data streams. Intell. Interact. Multimedia Syst. Serv. Smart Innovation Syst. Technol. **14**, 53–62 (2012)

98. L. Yan, F.H. Ma, Application of cubic exponential smoothing method to city underground deformation prediction. Technol. Econ. Areas Commun. **43**(5), 62–71 (2007)

99. Y. Li, F. Jia, Application of dynamic cubic exponential smoothing method to the application of predicting gdp of liaoning province. Silicon Valley **13**, 152–153 (2009)

100. T.M. Williams, Adaptive holt-winters forecasting. J. Oper. Res. Soc. **38**, 553–560 (1987)

101. W.J. Taylor, Smooth transition exponential smoothing. J. Forecast. **23**(6), 385–404 (2004)

102. W.J. Taylor, Volatility forecasting with smooth transition exponential smoothing. Int. J. Forecast. **20**(2), 273–286 (2004)

103. Y.Z. Ma, J. Rao, W.S. Hu, X.F. Meng, X. Han, Y. Zhang, Y.P. Chai, C.Q. Liu, An efficient index for massive iot data in cloud environment. in *Proceedings of the 21st ACM International Conference on Information and Knowledge Management, CIKM '12* (ACM, New York, 2012), pp. 2129–2133

104. P. Cudre-Mauroux, E. Wu, S. Madden, Trajstore: an adaptive storage system for very large trajectory data sets. in *Proceedings of the 2010 IEEE 26th International Conference on Data Engineering* (IEEE, 2010), pp. 109–120

105. K.Y. Whang, R. Krishnamurthy, *The Multilevel Grid File: A Dynamic Hierarchical Multidimensional File Structure* (Korea Advanced Institute of Science and Technology, Center for Artificial Intelligence Research, 1991)

106. F. Chang, J. Dean, S. Ghemawat, W.C. Hsieh, D.A. Wallach, M. Burrows, T. Chandra, A. Fikes, R.E. Gruber, Bigtable: A distributed storage system for structured data. ACM Trans. Database Syst. **26**(2), 4:1–4:26 (2008)

107. The Apache Software Foundation. Big data.http://hbase.apache.org/ (2013)

108. S. Nishimura, S. Das, D. Agrawal, A.E. Abbadi, Md-hbase: A scalable multi-dimensional data infrastructure for location aware services. in *Proceedings of the 2011 IEEE 12th International Conference on Mobile Data Management, MDM '11*, vol. 1 (IEEE Computer Society, Washington, 2011), pp. 7–16

109. H.Y. Tan, W.M. Luo, L.M. Ni, Clost: a hadoop-based storage system for big spatio-temporal data analytics. in *Proceedings of the 21st ACM International Conference on Information and knowledge Management, CIKM '12* (ACM, New York, 2012), pp. 2139–2143

110. L.L. Ding, B.Y. Qiao, G.R. Wang, C. Chen, An efficient quad-tree based index structure for cloud data management. in *Proceedings of the 12th International Conference on Web-Age Information Management, WAIM'11* (Springer, Heidelberg, 2011), pp. 238–250

111. J.B. Wang, S. Wu, H. Gao, J.Z. Li, B.C. Ooi, Indexing multi-dimensional data in a cloud system. in *Proceedings of the 2010 ACM SIGMOD International Conference on Management of data, SIGMOD '10* (ACM, New York, 2010), pp. 591–602

112. X.Y. Zhang, J. Ai, Z.Y. Wang, J.H. Lu, X.F. Meng, An efficient multi-dimensional index for cloud data management. in *Proceedings of the First International Workshop on Cloud Data Management, CloudDB '09* (ACM, New York, 2009), pp. 17–24

113. H.J. Liao, J.Z. Han, J.Y. Fang, Multi-dimensional index on hadoop distributed file system. in *Proceedings of the 2010 IEEE Fifth International Conference on Networking, Architecture and Storage (NAS)* (IEEE, 2010), pp. 240–249

114. J. Feng, C.Y. Lu, S.M. Xu, T. Watanabe, Dynsketch: a spatio-temporal aggregate index for moving objects in road networks. Int. J. Intell. Defence Support Syst. **2**(2), 120–137 (2009)-09-21T00:00:00

115. K.S. Candan, J.W.K.P. Nagarkar, M. Nagendra, R.W. Yu, Rankloud: Scalable multimedia data processing in server clusters. IEEE Multimedia **18**(1), 64–77 (2011)

116. C. Doulkeridis, K. Nørvåg, On saying "enough already!" in mapreduce. in *Proceedings of the 1st International Workshop on Cloud Intelligence, Cloud-I '12* (ACM, New York, 2012), pp. 7:1–7:4

117. W. Lu, Y.Y. Shen, S. Chen, B.C. Ooi, Efficient processing of k nearest neighbor joins using mapreduce. PVLDB **5**(10), 1016–1027 (2012)

118. L.L. Cao, J. Krumm, From gps traces to a routable road map. in *Proceedings of the 17th ACM SIGSPATIAL International Conference on Advances in Geographic Information Systems, GIS '09* (ACM, New York, 2009), pp. 3–12

119. Y.H. Chen, J. Krumm, Probabilistic modeling of traffic lanes from gps traces. in *Proceedings of the 18th SIGSPATIAL International Conference on Advances in Geographic Information Systems, GIS '10* (ACM, New York, 2010), pp. 81–88

120. S.Y. Liu, Y.H. Liu, L.M. Ni, J.P. Fan, M.L. Li, Towards mobility-based clustering. in *Proceedings of the 16th ACM SIGKDD International Conference on Knowledge Discovery and Data Mining, KDD '10* (ACM, New York, 2010), pp. 919–928

121. R.K. Balan, K.X. Nguyen, L.X. Jiang, Real-time trip information service for a large taxi fleet. in *Proceedings of the 9th International Conference on Mobile Systems, Applications, and Services, MobiSys '11* (ACM, New York, 2011), pp. 99–112

122. J. Yuan, Y. Zheng, X. Xie, G.Z. Sun, Driving with knowledge from the physical world. in *Proceedings of the 17th ACM SIGKDD International Conference on Knowledge Discovery and Data Mining, KDD '11* (ACM, New York, 2011), pp. 316–324

123. J. Yuan, Y. Zheng, C.Y. Zhang, W.L. Xie, X. Xie, G.Z. Sun, Y. Huang, T-drive: driving directions based on taxi trajectories. in *Proceedings of the 18th SIGSPATIAL International Conference on Advances in Geographic Information Systems, GIS '10* (ACM, New York, 2010), pp. 99–108

124. X.L. Li, G. Pan, Z.H. Wu, G.D. Qi, S.J. Li, D.Q. Zhang, W.S. Zhang, Z.H. Wang, Prediction of urban human mobility using large-scale taxi traces and its applications. Front. Comput. Sci China **6**(1), 111–121 (2012)

125. D.Q. Zhang, N. Li, Z.H. Zhou, C. Chen, L. Sun, S.J. Li, ibat: detecting anomalous taxi trajectories from gps traces. in *Proceedings of the 13th International Conference on Ubiquitous Computing, UbiComp '11* (ACM, New York, 2011), pp. 99–108

126. G. Pan, G. Qi, Z.H. Wu, D.Q. Zhang, S.J. Li, Land-use classification using taxi gps traces. IEEE Trans. Intell. Transp. Syst. **14**(1), 113–123 (2013)

127. Y Zheng, Y.C. Liu, J. Yuan, X. Xie, Urban computing with taxicabs. in *Proceedings of the 13th International Conference on Ubiquitous Computing, UbiComp '11* (ACM, New York, 2011), pp. 89–98

Printed in the United States
By Bookmasters